METHODS IN MOLECULAR BIOLOGY

Series Editor
John M. Walker
School of Life Sciences
University of Hertfordshire
Hatfield, Hertfordshire, AL10 9AB, UK

For further volumes:
http://www.springer.com/series/7651

Nuclear G-Protein Coupled Receptors

Methods and Protocols

Edited by

Bruce G. Allen

Montreal Heart Institute & Université de Montréal, Montreal, QC, Canada

Terence E. Hébert

Pharmacology and Therapeutics, McGill University, Montreal, QC, Canada

 Humana Press

Editors
Bruce G. Allen
Montreal Heart Institute & Université
  de Montréal
Montreal, QC, Canada

Terence E. Hébert
Pharmacology and Therapeutics
McGill University
Montreal, QC, Canada

ISSN 1064-3745   ISSN 1940-6029 (electronic)
ISBN 978-1-4939-1754-9   ISBN 978-1-4939-1755-6 (eBook)
DOI 10.1007/978-1-4939-1755-6
Springer New York Heidelberg Dordrecht London

Library of Congress Control Number: 2014949859

Printed on acid-free paper

Humana Press is a brand of Springer
Springer is part of Springer Science+Business Media (www.springer.com)

Preface

Signaling via Nuclear-Localized Transmembrane Receptors: New Homes and New Tasks for Cell Surface Proteins

Growth factor receptors and G protein-coupled receptors (GPCRs) are now known to be present on nuclear or perinuclear membranes, challenging a long-held paradigm regarding their sites of action being restricted to the cell surface membranes (reviewed in refs. 1–4). These observations have led to the concept of intracrine signaling. Intracrine signaling refers to processes whereby ligands, originating within a target cell or taken up from the extracellular milieu, act upon intracellular receptors (reviewed in refs. 3–11). Studies to date have localized Ang II type 1 (AT1R) and type 2 (AT2R) [12–14], bradykinin B2 (15), α- [16, 17] and β-adrenergic [18–20], type B endothelin (ETB) [21, 22], epidermal growth factor (EGF-R) [23], c-erbB-4 [24], insulin [25, 26], interferon β [27], muscarinic cholinergic [28], nerve growth factor [29], prostaglandin E_2 (PGE$_2$) [30, 31], lysophosphatidic acid type-1 [32], urotensin II [33, 34], and opioid [35] receptors to the nuclear membrane. Consistent with a nuclear location, several GPCRs have now been shown to contain a nuclear localization sequence (NLS) [17, 36–38]. At present, little is known concerning the function of nuclear growth factor receptors or GPCRs: potential roles include regulation of nuclear transport, transcription, and nuclear envelope formation.

Of course, growth factor receptors and GPCRs require the presence of a host of effector molecules in order to transduce signals. Key components of various signaling pathways are also present at the nuclear envelope or within the nucleus itself. These include heterotrimeric G proteins [13, 18, 39, 40] (reviewed in ref. 4), adenylyl cyclase [41], phosphodiesterase [42], diacylglycerol kinase-ζ [43], endothelial nitric oxide synthase (eNOS) [20], phospholipase A_2 [44], phospholipase C-β1 [45, 46], phospholipase D [47–49], phosphatidylinositol 3-kinase C2α [50], RGS proteins (reviewed in ref. 51), β-arrestin1 [52, 53], and G protein-coupled receptor kinases [54–56]. Nuclear membranes also contain sarco/endoplasmic Ca^{2+}-ATPase (SERCA) pumps [57, 58] as well as ryanodine-sensitive (RyR) [58, 59] and inositol trisphosphate-sensitive (IP$_3$R) [59, 60] Ca^{2+} channels. Hence, the nuclear cisternae are able to accumulate and release Ca^{2+} (reviewed in ref. 61). Furthermore, numerous protein kinases localize to the nucleus, or translocate into or out of the nucleus upon activation [62–66] and inhibition of ERK, p38α/β, PKB, PKG, or JNK inhibited both basal and/or isoproterenol-stimulated transcription in isolated nuclei [19, 20]. Thus, there is evidence supporting the presence and physiological relevance of GPCR signaling at the nuclear membrane.

For intracellular GPCRs to be of functional relevance, barring constitutive receptor activity, there must be a source of the intracellular ligand, whether it be taken up from the extracellular milieu or synthesized within the target cell (reviewed in ref. 3). Ligand-mediated receptor internalization and translocation to the nucleus have been demonstrated for AT1R [37] and EGF-R [67]. In contrast, PGE$_2$ and catecholamines are taken up by specific transporters [16, 68, 69]. When fluorescent ET-1 or ET-3 analogs were applied extracellularly, they were endocytosed and subsequently trafficked to lysosomes [22, 70], suggesting that endogenous endothelins may serve as ligands for nuclear ETB.

The nuclear envelope (NE) is a double-lipid bilayer structure, comprising inner (INM) and outer (ONM) nuclear membranes, that separates the nucleoplasm from the cytoplasm. The INM and ONM only meet at the nuclear pore: a large complex (>1,000 subunits) that facilitates the regulated exchange of macromolecules and RNA between the nucleoplasm and cytoplasm. The space between the INM and ONM is known as the nuclear cisternae or perinuclear space [71]. The ONM is contiguous with the sarco/endoplasmic reticulum whereas the INM is associated with the nucleoskeleton, a structure analogous to the cytoskeleton, that lines the inner surface of the INM. Furthermore, in adult cardiac myocytes, the sarcoplasmic reticulum and perinuclear space are interconnected [72]. By virtue of the linker of nucleoskeleton and cytoskeleton (LINC) complex, the nucleus is mechanically coupled to the cell surface and extracellular matrix (reviewed in ref. [73]). The LINC complex, comprising the SUN proteins (Sad1p/UNC-84) in the INM and the nesprins at the ONM, has been implicated both in nuclear morphology, positioning, and integrity and in mediating the effects of biomechanical load on gene expression [74]. Lamins A/C and B make up the nucleoskeleton and are coupled to the LINC complex via a direct interaction with SUN proteins. Furthermore, even isolated nuclei are capable of altering their stiffness in response to changes in mechanical force, or tension, and this is thought to involve a phosphorylation-dependent increase in coupling between lamins and the LINC complex [75]. However, it is also becoming increasingly apparent that the nuclear envelope itself is not just a barrier, but plays a role in signaling that may be as complex as that of the cell membrane. The objective of this installment in the series *Methods in Molecular Biology* is to highlight some of the methodologies used to identify and study signal process located in the nuclear membranes.

During their biosynthesis, as a result of an integral N-terminal signal motif, GPCRs orient within membranes with their effector binding site oriented towards the cytoplasm. Hence, whether a receptor is present on either the INM or the ONM, the ligand-binding site will likely be oriented towards the nuclear cisternae. It remains to be determined whether these receptors signal into either the cytoplasm, the nucleoplasm, or both. Hence, whether the receptor is on the INM, the ONM, or both, a significant issue remains—How would an intracrine ligand reach its binding site within the nuclear cisternae? The choices include (1) passive or active transit across the INM or ONM or (2) direct delivery to the nuclear cisternae during biosynthesis. This latter mechanism could include transport from the ER lumen to the nuclear cisternae: the SR and nuclear are interconnected in cardiac myocytes [72] and perhaps this connectivity permits protein trafficking. Alternatively, peptide ligands may arrive at the nucleus by microsomal transport. The nuclear envelope was recently shown to undergo budding [76]; hence, microsomes may also be able to fuse with the ONM and thus deliver their cargo into the perinuclear space.

Nuclear localization has been reported for many peptide ligands (e.g., Ang II, epidermal growth factor, insulin, platelet-derived growth factor, nerve growth factor, parathyroid hormone-related protein, prolactin, interleukin 1, somatostatin, fibroblast growth factors 1 and 2, and TGFα). Hence, the actual source of the intracrine ligand as well as the means whereby it is trafficked to its site of action within the nuclear cisternae may be distinct for each ligand, or class of ligands. Hence, intracrine signaling represents a potentially novel target for therapeutic intervention.

In this volume, we bring together a number of conceptual and methodological aspects important for the validation and characterization of intracrine signaling systems. To date, the best characterized intracrine signaling system is that of angiotensin II (Ang II). AT1R and AT2R have been demonstrated on the nuclear membranes in cardiomyocytes [3], hepatocytes, and vascular smooth muscle cell lines [13, 77]. In the first chapter of this book, Re and Cook recount the history of Ang II intracrine signaling. How to target such receptors to understand their physiology is the focus of the next several chapters. Internal receptors can be targeted by direct ligand injection into cells (Merlen and Ledoux, Chapter 2), and the use of caged ligands that can be released by UV irradiation to study nuclear GPCR signaling in an intact cell context (Chatenet et al., Chapter 3 and Merlen et al., Chapter 4). Exploring potential sources of ligands and measuring their uptake and delivery to the nucleus are the subject of Dahl et al., Chapter 5. Another interesting possibility for intracrine signaling is the association of the enzyme responsible for ligand synthesis with the receptor for the ligand in question as recently demonstrated for the prostaglandin D_2 DP1 receptor and the intracellular l-prostaglandin D synthase [78]. In this volume, Binda and Parent provide guidance as to how to identify and characterize such interactions with endogenous proteins (Chapter 6).

Methodology to study the subcellular localization and function of GPCRs and other signaling systems is provided in several chapters. Both Tadevosyan et al. (Chapter 7) and Bhosle et al. (Chapter 8) describe techniques to isolate nuclei and localize signaling molecules to this compartment. GPCRs, as discussed above, are not the only receptor family expressed on the nuclear membrane and biochemical techniques to further fractionate inner and outer nuclear membranes are critical tools to understand their trafficking and function (Wang et al., Chapter 9). Another critical issue for the study of intracrine signaling is the requirement to study these systems in native cells to avoid artifacts associated with overexpression. Jong and O'Malley describe methods to study nuclear GPCR signaling in neurons (Chapter 10) and the intact cardiomyocyte is the focus of a number of chapters (Merlin and Allen, Chapter 5; Ryall and Saucerman, Chapter 11; Ljuboyevic and Bers, Chapter 12; and Bossuyt and Bers, Chapter 13).

In addition, numerous chapters focus on methods designed to understand signaling mediated by nuclear and other internal GPCRs. Ryall and Saucerman (Chapter 5) use a high-content microscopy approach to examine phenotypic changes in neonatal cardiomyocytes, an approach that will be generally amenable to studying nuclear GPCR and RTK signaling in the intact cell context. Studying signaling in isolated nuclei has been the focus of many of the studies cited above. It has become clear that nuclear GPCRs control gene expression, and Vaniotis et al. (Chapter 14) describe methods to assess and validate how these receptors modulate transcription using transcription initiation assays, gene arrays, and qPCR. The connections between nuclear GPCRs, G proteins, and their effectors remain incompletely understood. It may be that signals from surface GPCRs and their associated G protein-dependent pathways may integrate with those from nuclear-localized signaling complexes as well as with receptor-independent G protein signaling. Thus, Campden et al. (Chapter 15) demonstrate an unbiased method of identifying nuclear G protein-interacting proteins in order to begin to characterize these integrated networks. As discussed, the nuclear membrane is not the only endomembrane site where GPCR signaling occurs, as described by Calebiro et al. (Chapter 16). Finally, methods are described to study the formation of second messengers such as cAMP and to study the trafficking of receptors from the cell surface (Calebiro et al., Chapter 16). Together, we incorporate a number of state-of-the-art approaches to characterize what is becoming a common theme in cellular signaling.

Montreal, QC, Canada   *Bruce G. Allen*
                                                                            Terence E. Hébert

References

1. Boivin B, Vaniotis G, Allen BG et al (2008) G protein-coupled receptors in and on the cell nucleus: a new signalling paradigm? J Recept Signal Transduct Res 28:15–28
2. Gobeil F, Fortier A, Zhu T et al (2006) G-protein-coupled receptors signalling at the cell nucleus: an emerging paradigm. Can J Physiol Pharmacol 84:287–297
3. Tadevosyan A, Vaniotis G, Allen BG et al (2012) G protein-coupled receptor signalling in the cardiac nuclear membrane: evidence and possible roles in physiological and pathophysiological function. J Physiol 590:1313–1330
4. Vaniotis G, Allen BG, Hébert TE (2011) Nuclear GPCRs in cardiomyocytes: An insider's view of β-adrenergic receptor signaling. Am J Physiol Heart Circ Physiol 301:H1754–H1764
5. Bkaily G, Avedanian L, Al-Khoury J et al (2011) Nuclear membrane receptors for ET-1 in cardiovascular function. Am J Physiol Regul Integr Comp Physiol 300:R251–R263
6. Carey RM (2012) Functional intracellular renin-angiotensin systems: Potential for pathophysiology of disease. Am J Physiol Regul Integr Comp Physiol 302:R479–R481
7. Cook JL, Re RN (2012) Lessons from in vitro studies and a related intracellular angiotensin II transgenic mouse model. Am J Physiol Regul Integr Comp Physiol 302:R482–R493
8. Ellis B, Li XC, Miguel-Qin E et al (2012) Evidence for a functional intracellular angiotensin system in the proximal tubule of the kidney. Am J Physiol Regul Integr Comp Physiol 302:R494–R509
9. Gwathmey TM, Alzayadneh EM, Pendergrass KD et al (2012) Novel roles of nuclear angiotensin receptors and signaling mechanisms. Am J Physiol Regul Integr Comp Physiol 302:R518–R530
10. Re RN, Cook JL (2011) Noncanonical intracrine action. J Am Soc Hypertens 5:435–448
11. Zhuo JL, Li XC (2011) New insights and perspectives on intrarenal renin-angiotensin system: focus on intracrine/intracellular angiotensin II. Peptides 32:1551–1565
12. Re RN, Vizard DL, Brown J et al (1984) Angiotensin II receptors in chromatin. J Hypertension 2:S271–S273
13. Booz GW, Conrad KM, Hess AL et al (1992) Angiotensin-II binding sites on hepatocyte nuclei. Endocrinology 130:3641–3649
14. Tadevosyan A, Maguy A, Villeneuve LR et al (2010) Nuclear-delimited angiotensin receptor-mediated signaling regulates cardiomyocyte gene expression. J Biol Chem 295:22338–22349
15. Savard M, Barbaz D, Belanger S et al (2008) Expression of endogenous nuclear bradykinin B2 receptors mediating signaling in immediate early gene activation. J Cell Physiol 216:234–244
16. Wright CD, Chen Q, Baye NL et al (2008) Nuclear α1-adrenergic receptors signal activated erk localization to caveolae in adult cardiac myocytes. Circ Res 103:992–1000
17. Wright CD, Wu SC, Dahl EF et al (2012) Nuclear localization drives α1-adrenergic receptor oligomerization and signaling in cardiac myocytes. Cell Signal 24:794–802
18. Boivin B, Lavoie C, Vaniotis G et al (2006) Functional β-adrenergic receptor signalling on nuclear membranes in adult rat and mouse ventricular cardiomyocytes. Cardiovasc Res 71:69–78
19. Vaniotis G, Del Duca D, Trieu P et al (2011) Nuclear β-adrenergic receptors modulate gene expression in adult rat heart. Cell Signal 23:89–98
20. Vaniotis G, Glazkova I, Merlen C et al (2013) Regulation of cardiac nitric oxide signalling by nuclear β-adrenergic and endothelin receptors. J Mol Cell Cardiol 62:58–68
21. Boivin B, Chevalier D, Villeneuve LR et al (2003) Functional endothelin receptors are present on nuclei in cardiac ventricular myocytes. J Biol Chem 278:29153–29163
22. Merlen C, Farhat N, Luo X et al (2013) Intracrine endothelin signaling evokes 3-dependent increases in nucleoplasmic Ca^{2+} in adult cardiac myocytes. J Mol Cell Cardiol 62:189–202
23. Carpentier IL, Rees AR, Gregoriou M et al (1986) Subcellular distribution of the external and internal domains of the EGF receptor in A-431 cells. Exp Cell Res 166:312–326
24. Srinivasan R, Gillett CE, Barnes DM et al (2000) Nuclear expression of the c-erbB-4/HER-4 growth factor receptor in invasive breast cancer. Cancer Res 60:1483–1487
25. Horvat A (1978) Insulin binding sites on rat liver nuclear membranes: Biochemical and immunological studies. J Cell Physiol 97:37–47
26. Vigneri R, Goldfine ID, Wong KY et al (1978) The nuclear envelope. The major site of insulin binding in rat liver nuclei. J Biol Chem 253:2098–2103
27. Kushnaryov VM, MacDonald HS, Sedmak JJ et al (1985) Murine interferon-beta receptor-mediated endocytosis and nuclear membrane binding. Proc Natl Acad Sci U S A 82:3281–3285
28. Lind GJ, Cavanagh HD (1993) Nuclear muscarinic acetylcholine receptors in corneal cells from rabbit. Invest Ophtalmol Visual Sci 34:2943–2952
29. Yankner BA, Shooter EM (1979) Nerve growth factor in the nucleus: Interaction with

receptors on the nuclear membrane. Proc Natl Acad Sci U S A 76:1269–1273

30. Bhattacharya M, Peri KG, Almazan G et al (1998) Nuclear localization of prostaglandin E2 receptors. Proc Natl Acad Sci U S A 95:15792–15797

31. Bhattacharya M, Peri K, Ribeiro-da-Silva A et al (1999) Localization of functional prostaglandin E2 receptors EP3 and EP4 in the nuclear envelope. J Biol Chem 274: 15719–15724

32. Gobeil F Jr., Bernier SG, Vazquez-Tello A et al (2003) Modulation of pro-inflammatory gene expression by nuclear lysophosphatidic acid receptor type-1. J Biol Chem 278:38875–38883

33. Doan ND, Nguyen TT, Letourneau M et al (2012) Biochemical and pharmacological characterization of nuclear urotensin-II binding sites in rat heart. Br J Pharmacol 166:243–257

34. Nguyen TT, Letourneau M, Chatenet D et al (2012) Presence of urotensin-II receptors at the cell nucleus: Specific tissue distribution and hypoxia-induced modulation. Int J Biochem Cell Biol 44:639–647

35. Belcheva M, Barg J, Rowinski J et al (1993) Novel opioid binding sites associated with nuclei of NG-108-15 neurohybrid cells. J Neurosci 13:104–114

36. Lee DK, Lanca AJ, Cheng R et al (2004) Agonist-independent nuclear localization of the apelin, angiotensin AT1, and bradykinin B2 receptors. J Biol Chem 279:7901–7908

37. Lu D, Yang H, Shaw G et al (1998) Angiotensin ii-induced nuclear targeting of the angiotensin type I (AT1) receptor in brain cells. Endocrinology 139:365–375

38. Pickard BW, Watson PH (2008) Nuclear trafficking of the G-protein-coupled parathyroid hormone receptor. Crit Rev Eukaryot Gene Expr 18:151–161

39. Zhang JH, Barr VA, Mo Y et al (2001) Nuclear localization of G protein β5 and regulator of G protein signaling 7 in neurons and brain. J Biol Chem 276:10284–10289

40. Boivin B, Villeneuve LR, Farhat N et al (2005) Subcellular distribution of endothelin signalling pathway components in ventricular myocytes and heart: Lack of preformed caveolar signalosomes. J Mol Cell Cardiol 38:665–676

41. Yamamoto S, Kawamura K, James TM (1998) Intracellular distribution of adenylate cyclase in human cardiocytes determined by electron microscopic cytochemistry. Microsc Res Tech 40:479–498

42. Lugnier C, Keravis T, Le Bec A et al (1999) Characterization of cyclic nucleotide phosphodiesterase isoforms associated to isolated cardiac nuclei. Biochim Biophys Acta 1472:431–446

43. Topham MK, Bunting M, Zimmerman GA et al (1998) Protein kinase C regulates the nuclear localization of diacylglycerol kinase-ζ. Nature (London) 394:697–700

44. Fatima S, Yaghini FA, Ahmed A et al (2003) Cam kinase iia mediates norepinephrine-induced translocation of cytosolic phospholipase A₂ to the nuclear envelope. J Cell Sci 116:353–365

45. Kim CG, Park D, Rhee SG (1996) The role of carboxyl-terminal basic amino acids in gqα-dependent activation, particulate associate, and nuclear localization of phospholipase C-β1. J Biol Chem 271:21187–21192

46. Faenza I, Matteucci A, Manzoli L et al (2000) A role for nuclear phospholipase Cβ₁ in cell cycle control. J Biol Chem 275:30520–30524

47. Baldassare JJ, Jarpe MB, Alferes L et al (1997) Nuclear translocation of RhoA mediates the mitogen-induced activation of phospholipase D involved in nuclear envelope signal transduction. J Biol Chem 272:4911–4914

48. Freyberg Z, Sweeney D, Siddhanta A et al (2001) Intracellular localization of phospholipase D1 in mammalian cells. Mol Biol Cell 12:943–955

49. Gayral S, Déléris P, Laulagnier K et al (2006) Selective activation of nuclear phospholipase D-1 by G protein-coupled receptor agonists in vascular smooth muscle cells. Circ Res 99:132–139

50. Didichenko SA, Thelen M (2001) Phosphatidylinositol 3-kinase c2α contains a nuclear localization sequence and associates with nuclear speckles. J Biol Chem 276:48135–48142

51. Burchett SA (2003) In through the out door: nuclear localization of the regulators of G protein signaling. J Neurochem 87:551–559

52. Scott MG, Le Rouzic E, Perianin A et al (2002) Differential nucleocytoplasmic shuttling of β-arrestins. Characterization of a leucine-rich nuclear export signal in β-arrestin2. J Biol Chem 277:37693–37701

53. Wang P, Wu Y, Ge X et al (2003) Subcellular localization of β-arrestins is determined by their intact N domain and the nuclear export signal in the C terminus. J Biol Chem 278:11648–11653

54. Yi XP, Gerdes AM, Li F (2002) Myocyte redistribution of GRK2 and GRK5 in hypertensive, heart-failure-prone rats. Hypertension 39:1058–1063

55. Yi XP, Zhou J, Baker J et al (2005) Myocardial expression and redistribution of GRKs in hypertensive hypertrophy and failure. Anat Rec A Discov Mol Cell Evol Biol 282:13–23

56. Johnson LR, Scott MG, Pitcher JA (2004) G protein-coupled receptor kinase 5 contains a DNA-binding nuclear localization sequence. Mol Cell Biol 24:10169–10179

57. Lanini L, Bachs O, Carafoli E (1992) The calcium pump of liver nuclear membrane is

identical to that of endoplasmic reticulum. J Biol Chem 267:11548–11552

58. Abrenica B, Gilchrist JSC (2000) Nucleoplasmic Ca²⁺ loading is regulated by mobilization of perinuclear Ca²⁺. Cell Calcium 28:127–136

59. Guihard G, Proteau S, Rousseau E (1997) Does the nuclear envelope contain two types of ligand-gated Ca²⁺ release channels? FEBS Lett 414:89–94

60. Bare DJ, Kettlun CS., Liang M et al (2005) Cardiac type 2 inositol 1,4,5-trisphosphate receptor: Interaction and modulation by calcium/calmodulin-dependent protein kinase II. J Biol Chem 280:15912–15920

61. Bootman MD, Fearnley C, Smyrnias I et al (2009) An update on nuclear calcium signalling. J Cell Sci 122:2337–2350

62. Simonson MS, Herman WH (1993) Protein kinase C and protein tyrosine kinase activity contribute to mitogenic signaling by endothelin-1. Cross-talk between G protein-coupled receptors and pp60ᶜ⁻ˢʳᶜ. J Biol Chem 268:9347–9357

63. Sadoshima J, Qiu Z, Morgan JP et al (1995) Angiotensin II and other hypertrophic stimuli mediated by G protein-coupled receptors activate tyrosine kinase, mitogen-activated protein kinase, and 90-kD S6 kinase in cardiac myocytes. The critical role of Ca²⁺-dependent signaling. Circ Res 76:1–15

64. Shubeita HE, McDonough PM, Harris AN et al (1990) Endothelin induction of inositol phospholipid hydrolysis, sarcomere assembly, and cardiac gene expression in ventricular myocytes. A paracrine mechanism for myocardial cell hypertrophy. J Biol Chem 265:20555–20562

65. Bogoyevitch MA, Glennon PE, Sugden PH (1993) Endothelin-1, phorbol esters and phenylephrine stimulate map kinase activities in ventricular cardiomyocytes. FEBS Lett 317:271–275

66. Bogoyevitch MA, Glennon PE, Andersson MB et al (1994) Endothelin-1 and fibroblast growth factors stimulate the mitogen-activated protein kinase signalling cascade in cardiac myocytes. The potential role of the cascade on the integration of two signalling pathways leading to myocyte hypertrophy. J Biol Chem 269:1110–1119

67. Marti U, Ruchti C, Kampf J et al (2001) Nuclear localization of epidermal growth factor and epidermal growth factor receptors in human thyroid tissues. Thyroid 11:137–145

68. Gobeil F Jr, Dumont I, Marrache AM et al (2002) Regulation of enos expression in brain endothelial cells by perinuclear EP₃ receptors. Circ Res 90:682–689

69. Ryall KA, Saucerman JJ (2012) Automated imaging reveals a concentration dependent delay in reversibility of cardiac myocyte hypertrophy. J Mol Cell Cardiol 53:282–290

70. Oksche A, Boese G, Horstmeyer A et al (2000) Late endosomal/lysosomal targeting and lack of recycling of the ligand-occupied endothelin B receptor. Mol Pharmacol 57:1104–1113

71. Stehno-Bittel L, Perez-Terzic C, Clapham DE (1995) Diffusion across the nuclear envelope inhibited by depletion of the nuclear Ca²⁺ store. Science 270:1835–1838

72. Wu X, Bers DM (2006) Sarcoplasmic reticulum and nuclear envelope are one highly interconnected Ca²⁺ store throughout cardiac myocyte. Circ Res 99:283–291

73. Stroud MJ, Banerjee I, Veevers J et al (2014) Linker of nucleoskeleton and cytoskeleton complex proteins in cardiac structure, function, and disease. Circ Res 114: 538–548

74. Banerjee I, Zhang J, Moore-Morris T et al (2014) Targeted ablation of nesprin 1 and nesprin 2 from murine myocardium results in cardiomyopathy, altered nuclear morphology and inhibition of the biomechanical gene response. PLoS Genet 10:e1004114

75. Guilluy C, Osborne LD, Van Landeghem L et al (2014) Isolated nuclei adapt to force and reveal a mechanotransduction pathway in the nucleus. Nat Cell Biol doi:10.1038/ncb2927

76. Speese SD, Ashley J, Jokhi V et al (2012) Nuclear envelope budding enables large ribonucleoprotein particle export during synaptic wnt signaling. Cell 149:832–846

77. Cook JL, Mills SJ, Naquin R et al (2006) Nuclear accumulation of the AT1 receptor in a rat vascular smooth muscle cell line: effects upon signal transduction and cellular proliferation. J Mol Cell Cardiol 40:696–707

78. Binda C, Genier S, Cartier A et al (2014) A G protein-coupled receptor and the intracellular synthase of its agonist functionally cooperate. J Cell Biol 204:377–393

Contents

Contributors

BRUCE G. ALLEN • *Montreal Heart Institute, Montréal, QC, Canada; Department of Medicine, Université de Montréal, Montréal, QC, Canada; Department of Biochemistry and Molecular Medicine, Université de Montréal, Montréal, QC, Canada; Department of Pharmacology and Therapeutics, McGill University, Montréal, QC, Canada*

STÉPHANE ANGERS • *Leslie Dan Faculty of Pharmacy, University of Toronto, Toronto, ON, Canada*

NICOLAS AUDET • *Department of Pharmacology and Therapeutics, McGill University, Montréal, QC, Canada*

DONALD M. BERS • *Department of Pharmacology, University of California, Davis, CA, USA*

VIKRANT K. BHOSLE • *Department of Pharmacology and Therapeutics, McGill University, Montreal, QC, Canada; CHU Sainte-Justine Hospital Research Centre, Montreal, QC, Canada; Research Centre of Maisonneuve-Rosemont Hospital, Montreal, QC, Canada*

CHANTAL BINDA • *Department of Medicine, Université de Sherbrooke, Sherbrooke, QC, Canada*

JULIE BOSSUYT • *Department of Pharmacology, University of California, Davis, CA, USA*

STEVE BOURGAULT • *Department of Chemistry, University of Québec in Montreal, Montreal, QC, Canada*

DAVIDE CALEBIRO • *Bio-Imaging Center/Rudolf Virchow Center for Experimental Biomedicine, and Institute of Pharmacology and Toxicology, University of Würzburg, Würzburg, Germany*

RHIANNON CAMPDEN • *Department of Pharmacology and Therapeutics, McGill University, Montréal, QC, Canada*

DAVID CHATENET • *INRS – Institut Armand-Frappier, Institut National de la Recherche Scientifique, QC, Canada*

SYLVAIN CHEMTOB • *Department of Pharmacology and Therapeutics, McGill University, Montreal, QC, Canada; Departments of Pediatrics, Ophthalmology and Pharmacology, CHU Sainte-Justine Hospital Research Centre, Montreal, QC, Canada; Research Centre of Maisonneuve-Rosemont Hospital, Montreal, QC, Canada*

JULIA L. COOK • *Department of Cardiology, Ochsner Health System, Tulane University School of Medicine, New Orleans, LA, USA; Ochsner Health System, Institute for Clinical Research, Louisiana State University Health Sciences Center, New Orleans, LA, USA; Department of Biochemistry and Molecular Biology, Louisiana State University Health Sciences Center, New Orleans, LA, USA; Neuroscience Center of Excellence, Louisiana State University Health Sciences Center, New Orleans, LA, USA*

ERIKA F. DAHL • *Department of Pharmacology, University of Minnesota, Minneapolis, MN, USA*

ALAIN FOURNIER • *INRS – Institut Armand-Frappier, Institut National de la Recherche Scientifique, QC, Canada*

FERNAND GOBEIL JR • *Department of Pharmacology, Université de Sherbrooke, Sherbrooke, QC, Canada*

AMOD GODBOLE • *Bio-Imaging Center/Rudolf Virchow Center for Experimental Biomedicine, and Institute of Pharmacology and Toxicology, University of Würzburg, Würzburg, Germany*

SARAH GORA • *Department of Pharmacology and Therapeutics, McGill University, Montréal, QC, Canada*

TERENCE E. HÉBERT • *Department of Pharmacology and Therapeutics, McGill University, Montreal, QC, Canada*

JENNIFER L. HSU • *Department of Molecular and Cellular Oncology, The University of Texas MD Anderson Cancer Center, Houston, TX, USA; Center for Molecular Medicine and Graduate Institute of Cancer Biology, China Medical University, Taichung, Taiwan*

MIEN-CHIE HUNG • *Department of Molecular and Cellular Oncology, The University of Texas MD Anderson Cancer Center, Houston, TX, USA; Center for Molecular Medicine and Graduate Institute of Cancer Biology, China Medical University, Taichung, Taiwan*

LONGFEI HUO • *Department of Molecular and Cellular Oncology, The University of Texas MD Anderson Cancer Center, Houston, TX, USA*

YUH-JIIN I. JONG • *Department of Anatomy and Neurobiology, Washington University School of Medicine, Saint Louis, MO, USA*

JONATHAN LEDOUX • *Montreal Heart Institute, Montreal, QC, Canada; Department of Medicine, Université de Montréal, Montreal, QC, Canada; Department of Molecular and Integrative Physiology, Université de Montréal, Montreal, QC, Canada*

SENKA LJUBOJEVIĆ • *Department of Pharmacology, University of California Davis, Davis, CA, USA; Department of Cardiology, Medical University of Graz, Graz, Austria*

MARTIN J. LOHSE • *Bio-Imaging Center/Rudolf Virchow Center for Experimental Biomedicine, and Institute of Pharmacology and Toxicology,, University of Würzburg, Würzburg, Germany*

SANDRA LYGA • *Bio-Imaging Center/Rudolf Virchow Center for Experimental Biomedicine, and Institute of Pharmacology and Toxicology,, University of Würzburg, Würzburg, Germany*

CLÉMENCE MERLEN • *Montreal Heart Institute, Montreal, QC, Canada; Department of Biochemistry and Molecular Medicine Université de Montréal, Montreal, QC, Canada*

ANDRÉ NANTEL • *National Research Council Canada, Montreal, QC, Canada; Department of Anatomy and Cell Biology, McGill University, Montreal, QC, Canada*

STANLEY NATTEL • *Montreal Heart Institute, Montreal, QC, Canada; Department of Medicine, Université de Montréal, Montreal, QC, Canada; Department of Pharmacology and Therapeutics, McGill University, Montreal, QC, Canada*

TIMOTHY D. O'CONNELL • *Department of Integrative Biology and Physiology, University of Minnesota, Minneapolis, MN, USA*

KAREN L. O'MALLEY • *Department of Anatomy and Neurobiology, Washington University School of Medicine, Saint Louis, MO, USA*

JEAN-LUC PARENT • *Department of Medicine, Université de Sherbrooke, Sherbrooke, QC, Canada*

DARLAINE PÉTRIN • *Department of Pharmacology and Therapeutics, McGill University, Montréal, QC, Canada*

RICHARD N. RE • *Departments of Medicine and Physiology, Tulane University School of Medicine, New Orleans, LA, USA; Department of Cardiology, Ochsner Health System, Tulane University School of Medicine, New Orleans, LA, USA*

ALFREDO RIBEIRO-DA-SILVA • *Department of Pharmacology and Therapeutics, McGill University, Montreal, QC, Canada*

JOSE CARLOS RIVERA • *CHU Sainte-Justine Hospital Research Centre, Montreal, QC, Canada; Research Centre of Maisonneuve-Rosemont Hospital, Montreal, QC, Canada*

MÉLANIE ROBITAILLE • *Leslie Dan Faculty of Pharmacy, University of Toronto, Toronto, ON, Canada*

KAREN A. RYALL • *Department of Biomedical Engineering, University of Virginia, Charlottesville, VA, USA*

JEFFREY J. SAUCERMAN • *Department of Biomedical Engineering, University of Virginia, Charlottesville, VA, USA*

ARTAVAZD TADEVOSYAN • *Montreal Heart Institute, Montreal, QC, Canada; Department of Medicine, Université de Montréal, Montreal, QC, Canada*

GEORGE VANIOTIS • *Montreal Heart Institute, Montreal, QC, Canada; Department of Biochemistry and Molecular Medicine, Université de Montréal, Montreal, QC, Canada*

LOUIS R. VILLENEUVE • *Montreal Heart Institute, Montreal, QC, Canada*

YING-NAI WANG • *Department of Molecular and Cellular Oncology, The University of Texas MD Anderson Cancer Center, Houston, TX, USA; Center for Molecular Medicine and Graduate Institute of Cancer Biology, China Medical University, Taichung, Taiwan*

CASEY D. WRIGHT • *Department of Integrative Biology and Physiology, University of Minnesota, Minneapolis, MN, USA*

Chapter 1

Studies of Intracellular Angiotensin II

Richard N. Re and Julia L. Cook

Abstract

Many extracellular signaling proteins act within their cells of synthesis and/or in target cells after internalization. This type of action is called intracrine and it plays a role in diverse biological processes. The mechanisms of intracrine intracellular action are becoming clear thanks to the application of modern techniques of molecular biology. Here, progress in this area is reviewed. In particular the intracrine biology of angiotensin II is discussed.

Key words Angiotensin II, Intracrine, Growth factors

1 Introduction

The earliest studies of nuclear angiotensin binding to nuclei involved the injection of tritiated angiotensin II into animals, followed by electron microscopic autoradiography. These studies detected angiotensin II trafficking to nuclei and led to a wide range of follow-up studies [2–26]. Binding studies were conducted on isolated nuclei, and transcriptional runoff studies defined the positive effect of nuclear angiotensin II binding on transcription. These techniques were refined over time to the point that both AT 1 and AT-2 receptors (AT-1R and AT-2R) could be shown to be on the nuclear membrane and AT-1R-like receptors were associated with chromatin. Direct binding of angiotensin II to isolated nuclei was shown to up-regulate genes such as renin, angiotensinogen, and platelet-derived growth factor (PDGF). Collectively, these early studies revealed the existence of physiologically relevant angiotensin II action at nucleus, although the full scope of that activity was, and remains, only partially understood [1–9, 16, 20, 25, 26].

In time, the development of more precise techniques confirmed and expanded these early observations. For example, improved electron microscopy immuno-staining techniques confirmed angiotensin II binding to euchromatin. The transfection of cells with AT-1R fluorescent fusion proteins, coupled with confocal or

Bruce G. Allen and Terence E. Hébert (eds.), *Nuclear G-Protein Coupled Receptors*, Methods in Molecular Biology, vol. 1234, DOI 10.1007/978-1-4939-1755-6_1, © Springer Science+Business Media New York 2015

3D deconvolution microscopy, demonstrated the trafficking of the receptor from cell membrane to nucleus after exposure to extracellular ligand. Transfection of cells with a variety of constructs designed to generate angiotensin II solely in the intracellular space informed our view of the actions of the peptide within the cell and the establishment of novel transgenic mouse models now offers the prospect of new breakthroughs in the near future [8–25]. Below, we describe some of the models we have developed and review their implications.

2   Intracellular Angiotensin II

In order to further explore the actions of intracellular angiotensin II (iAII), we developed a construct designed to lead to the generation of iAII. Specifically, we mutated angiotensinogen cDNA by removing the region that encodes the signal-sequence [Ang(-S) Exp]. This was intended to produce an angiotensinogen which remained in the intracellular space. A corresponding control construct was also developed. Of note, both constructs contained a Kozak consensus sequence to enhance ribosome association and therefore translation. Rat hepatoma cells were then stably transfected with [Ang(-S)EXP] and control expression plasmids. These cells expressed renin as well as angiotensin-converting enzyme. We were able to show that these cells produced a previously described non-secreted form of renin which was expected to remain in the intracellular space in an active form, as opposed to pro-renin. Expression of [Ang(-S)Exp], but not control expression plasmid, resulted in increased proliferation. Anti-angiotensin II antibodies delivered to the culture medium did not affect this stimulation of mitosis. Proliferation was blocked by renin antisense and the angiotensin receptor blocker (ARB) losartan which had been shown to be internalized after binding to the cell surface AT-1R. The ARB candesartan, which fixes AT-1R at the cell surface, did not block the proliferative effect of [Ang(-S)EXP]. Phenylarsine oxide, which blocks receptor internalization, prevented the antiproliferative effect of losartan. We also found that [Ang(-S)Exp] up-regulated platelet-derived growth factor expression and the addition of anti-PDGF antibody to the culture medium partially blocked proliferation. These results argue for a direct effect of iAII on proliferation and PDGF up-regulation. Inhibition of intracellular iAII synthesis or blockade of intracellular AT-1R blocked this proliferation. Collectively, these results argued that the intracellular generation of iAII can occur with physiologically relevant results and that at least some of the effects of iAII are AT-1R mediated [8, 9].

In order to more directly investigate the trafficking of iAII, AT1R, and the effects of their interaction, we studied A10 rat vascular smooth muscle cells transfected with both a plasmid encoding a fluorescent AII fusion protein (pAII/ECFP encoding AII fused downstream of enhanced cyan fluorescent protein) and a plasmid encoding a rat AT-1Ra fluorescent fusion protein (pAT1R/EYFP, encoding the AT-1Ra receptor upstream of enhanced yellow fluorescent protein). The AII fusion protein did not contain a secretion signal and 3D deconvolution microscopy revealed its localization in the cell nuclei. AII was not detected in the media of transfected cells. AT-1Ra, when expressed in the absence of AII/ECFP, was localized to cytoplasm and cell surface and was not found in cell nucleus. However, when AII was introduced into the culture medium the receptor fusion protein localized to nucleus, just as it did when the cells were simultaneously transfected with pAT1R/EYFP and pAII/ECFP. In doubly transfected cells, both AT-1Ra and AII fusion proteins were found throughout the nucleus, although the receptor localized in nucleoli to a larger extent than did the AII fusion protein Transfection of the cells with pAII/ECFP alone resulted in increased proliferation as did the introduction of AII into the culture medium. However, while anti-angiotensin II antibody, delivered to the culture media, prevented the proliferation induced by extracellular AII, it had no effect on the proliferation stimulated by iAII. Collectively, these studies confirmed a role for iAII in the regulation of cellular proliferation and the interaction of iAII with AT1R in the intracellular space. Co-trafficking of AT1Ra and either extracellular or intracellular AII to nucleus was consistent with earlier autoradiographic studies of AII trafficking from the extracellular space to nucleus and with studies of isolated nuclei demonstrating widely dispersed nuclear receptors and effects of nuclear AII on gene regulation [10, 11].

Further studies of the actions of iAII in this model revealed that extracellular AII treatment of A10 cells activated cAMP response element-binding protein (CREB) as assessed by one-hybrid assays and immunoblotting. Expression of iAII similarly increased CREB but via different signaling pathways. Western blot and one-hybrid assays revealed that exogenous AII activated p38 MAPK and ERK 1/2. This finding was confirmed by demonstrating that the p38MAPK inhibitor SB203580 and the MEK inhibitor PD98059 each partially blocked the activation of CREB by exogenous AII. However, intracellular AII/ECFP activated p38MAPK in these cells but not ERK 1/2. Possibly the shared signaling pathway represented signaling by the internalized membrane receptor or AII binding to intracellular receptors, while the pathway exclusively activated by extracellular AII represents signaling by receptor on the cell surface [10, 11]. The details of intracellular AII action remain to be completely worked out [22].

3 Transgenic Mice

In order to shed light on the physiological effects of iAII we elected to employ transgenic technology. With our colleagues at the University of Iowa, we developed a line carrying the AII/ECFP construct regulated by the metallothionine promoter [12, 13]. Although the expression of the construct was widely distributed in tissues (and could be easily assessed using the attendant fluorescence), plasma angiotensin II concentrations were not elevated.

Although two independent mouse lines widely expressed the transgene (both RNA and protein), the prominent phenotypic change occurred in the kidney. By 2 months of age these mice developed systolic and diastolic hypertension. Some mice then developed glomerular thrombotic microangiopathy (TMA) involving microthrombosis of the glomerular capillaries and small vessels. This pathological picture has been reported in other transgenic models that over-express components of the renin-angiotensin-aldosterone system (RAAS). Blood pressure rose by 2 months of age while TMA developed at 4–6 months to varying degrees in individual animals. We found similar changes in males and females which contrast with some reports of sexually dimorphic blood pressure changes following the infusion of angiotensin II.

When isolated mouse embryonic fibroblasts (MEFs) from the transgenic animals were studied, ECFP/AII was detected in cytoplasm and in 30–40 % of nuclei. The cytoplasmic fluorescence was punctate and localized with mitochondrial markers. Multiple studies, including ours, have shown AII and/or AII receptors to be associated with mitochondria in adrenal, kidney, and brain. Moreover, mitochondrial dysfunction has been reported to play a role in RAAS-mediated disease and specifically in vascular endothelial dysfunction and hypertension, likely as the result of increasing levels of reactive oxygen species (ROS). In this regard, yeast two-hybrid studies followed by glutathione S-transferase (GST) pull-down studies revealed that AII associated with two mitochondrial electron transport chain proteins and increased ROS generation. Direct binding of AII to these proteins would preclude AII binding to integral membrane AT-1R. This raises the possibility that AII acts in the mitochondria in what we have termed a noncanonical fashion—that is, independent of membrane-associated receptor [12–14]. Recently, others using immunogold staining have reported AT-1R and AT-2R associated with mitochondria. The density of AT-1R per mitochondria increased with age while the AT-2R number fell. These authors further reported that mitochondrial AT-2R was coupled with nitric oxide production, thereby suggesting a canonical role for AII in mitochondria, that is, a function mediated by membrane-associated receptor [15]. It should be noted that a more recent study failed to identify AT-1aR in rat adrenal gland mitochondria while clearly identifying receptor in the nucleus [16]. The reason for the discordant mitochondrial findings is at this time unknown.

4   Canonical and Noncanonical Intracrine Action

For the purposes of this discussion we will deem signaling by membrane-bound receptor employing established second messenger cascades as canonical irrespective of whether the receptor resides in the cell surface membrane, the nuclear membrane, the endosome, the endolysosome, or elsewhere. Signaling that is not mediated by a membrane-associated receptor or that is independent of typical signaling mechanisms can be deemed noncanonical [14]. For example, the biological activity of the angiogenic factor angiogenin is dependent on internalization, nuclear trafficking, and RNase activity, the latter requirement rendering its action noncanonical. Some receptors may be activated at the cell surface, be internalized along with ligand, traffic to the nucleus, and then activate traditional second messengers even independent of membrane. This will be considered non-canonical because receptor-ligand trafficking to target responses to specific cell locations is reasonably considered atypical. Perhaps "quasi-canonical" would also be appropriate. Cleavage of the cytoplasmic tail of a cell surface receptor following ligand binding resulting in the trafficking of the fragment to nucleus or some other site whereupon it produces a distinct biological action is deemed non-intracrine because the ligand is no longer present. Of note in this regard, during our studies of AT 1R fluorescent fusion proteins we detected the cleavage of the receptor with the cytoplasmic fragment trafficking to nucleus [17]. Low levels of cleavage occurred in the absence of administered angiotensin II, but cleavage was markedly increased following the addition of angiotensin II. Using mass spectroscopy and Edman sequencing we determined that cleavage occurred between Leu(305) and Gly(306) at the junction of the seventh transmembrane domain and the cytoplasmic carboxy-terminal domain. Moreover, trafficking of the fragment to nucleus was associated in a variety of cell lines with caspase activation and apoptosis. Thus, although this action is not intracrine per se, it could well be biologically relevant in conditions such as heart failure, where angiotensin II is up-regulated and cardiac myocyte apoptosis is ongoing.

Canonical angiotensin action involving both AT-1R and AT-2R at the nuclear membrane has been established [14–16, 18]. In addition, receptors have been associated with (eu)chromatin and in particular with nucleosomes. Indeed, angiotensin II binding to these receptors induces conformational changes in chromatin consistent with gene regulation [4–7]. Given the absence of membrane lipid at the nucleosome, this action must be considered non-canonical or atypical. Thus taken together, the available data support both canonical and noncanonical angiotensin II action at both nucleus and mitochondria. This significantly expands our appreciation of the potential action of this peptide.

---

5 Conclusions and Future Directions

The study of the intracellular actions of angiotensin II and other intracrine ligands has progressed significantly over the last 30 years. In fact, there is much more to intracrine biology than is demonstrated by the biology of angiotensin II discussed here. Intracrine loops involve a very diverse group of proteins that often signal in the intracellular space in unanticipated ways. Although not discussed here, principles of intracrine biology have been proposed and these appear to have considerable heuristic value, not only in the understanding of angiotensin action but also in defining novel intracrine action in development, neoplasia, angiogenesis, and a variety of other physiological and pathological processes [19–25]. For example, in many systems intracrine signals display similar mechanisms of intercellular trafficking, and amplification in target cells through the establishment of feed-forward regulatory loops (Fig. 1). Angiotensin has been reported to follow this pattern, up-regulating angiotensinogen and renin after acting at the nuclei of diabetic cardiomyocytes and thereby creating a feed-forward loop in these cells [26]. Wider studies of the intracrine signals, their unexpected actions, and the general principles by which they act will likely serve to not only further inform our understanding of angiotensin II biology but also suggest new areas of investigation. A fuller appreciation of angiotensin II action could point the way to the development of drug candidates capable of interrupting canonical or noncanonical angiotensin II action with therapeutic benefits. But also, because intracrine biology is widely operative, it very likely encompasses many previously unrecognized and unsuspected protein actions. For example, the possible extension of the principles of intracrine biology to transmissible spongiform encephalopathies and other neurodegenerative disorders characterized by the cell-to-cell trafficking of prion-like proteins, by intracrine amplification, and by unexpected intracellular protein action could represent a novel, and potentially therapeutic, application of those principles [27].

In summary, intracrine biology, in large part developed through the study of angiotensin II, is supported by a multitude of observations and has made correct predictions about the actions of protein hormones. It has heuristic value in the near term and promises new insights into many biological processes in the future.

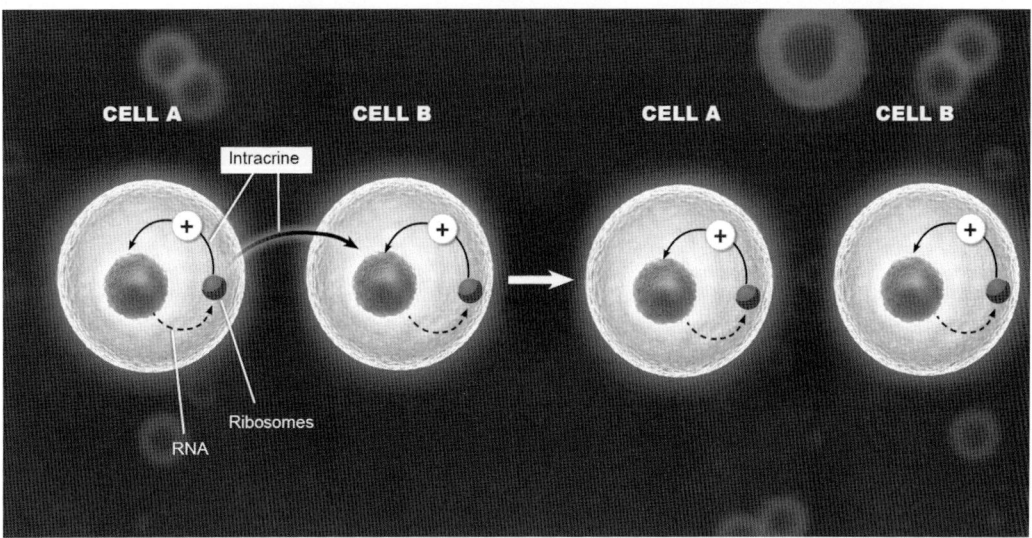

Fig. 1 Intracrines can act within their cells of synthesis (*cell A*) or after secretion and internalization by target cells (*cell B*). In either case, intracellular intracrines can establish self-sustaining feed-forward regulatory loops and thereby produce an active form of differentiation

References

1. Robertson AL Jr, Khairallah PA (1971) Angiotensin II: rapid localization in nuclei of smooth and cardiac muscle. Science 172: 1138–1139

2. Re RN, MacPhee AA, Fallon JT (1981) Specific nuclear binding of angiotensin II by rat liver and spleen nuclei. Clin Sci (Lond) 61:245S–247S

3. Re R, Parab M (1984) Effect of angiotensin II on RNA synthesis by isolated nuclei. Life Sci 34:647–651

4. Re RN, LaBiche RA, Bryan SE (1983) Nuclear-hormone mediated changes in chromatin solubility. Biochem Biophys Res Commun 110: 61–68

5. Re RN, Vizard DL, Brown J et al (1984) Angiotensin II receptors in chromatin fragments generated by micrococcal nuclease. Biochem Biophys Res Commun 119:220–227

6. Re RN, Vizard DL, Brown J et al (1984) Angiotensin II receptors in chromatin. J Hypertens Suppl 2:S271–S273

7. Re R, Bryan SE (1984) Functional intracellular renin-angiotensin systems may exist in multiple tissues. Clin Exp Hypertens A 6:1739–1742

8. Cook JL, Zhang Z, Re RN (2001) In vitro evidence for an intracellular site of angiotensin action. Circ Res 89:1138–1146

9. Cook JL, Giardina JF, Zhang Z et al (2002) Intracellular angiotensin II increases the long isoform of PDGF mRNA in rat hepatoma cells. J Mol Cell Cardiol 34:1525–1537

10. Cook JL, Mills SJ, Naquin R et al (2006) Nuclear accumulation of the AT1 receptor in a rat vascular smooth muscle cell line: effects upon signal transduction and cellular proliferation. J Mol Cell Cardiol 40:696–707

11. Cook JL, Mills SJ, Naquin RT et al (2007) Cleavage of the angiotensin II type 1 receptor and nuclear accumulation of the cytoplasmic carboxy-terminal fragment. Am J Physiol Cell Physiol 292:C1313–C1322

12. Redding KM, Chen BL, Singh A et al (2010) Transgenic mice expressing an intracellular fluorescent fusion of angiotensin II demonstrate renal thrombotic microangiopathy and elevated blood pressure. Am J Physiol Heart Circ Physiol 298:H1807–H1818

13. Cook JL, Re RN (2011) Lessons from in vitro studies and a related intracellular angiotensin II transgenic mouse model. Am J Physiol Regul Integr Comp Physiol 302:R482–R493

14. Re RN, Cook JL (2011) Noncanonical intracrine action. J Am Soc Hypertens 5:435–448

15. Abadir PM, Foster DB, Crow M et al (2011) Identification and characterization of a functional

mitochondrial angiotensin system. Proc Natl Acad Sci U S A 108:14849–14854

16. Premer C, Lamondin C, Mitzey A et al (2013) Immunohistochemical localization of AT1a, AT1b, and AT2 angiotensin II receptor subtypes in the rat adrenal, pituitary, and brain with a perspective commentary. Int J Hypertens 2013:175428. doi:10.1155/2013/175428

17. Cook JL, Singh A, DeHaro D et al (2011) Expression of a naturally occurring angiotensin AT(1) receptor cleavage fragment elicits caspase-activation and apoptosis. Am J Physiol Cell Physiol 301:C1175–C1185

18. Tadevosyan A, Vaniotis G, Allen BG et al (2012) G protein-coupled receptor signalling in the cardiac nuclear membrane: evidence and possible roles in physiological and pathophysiological function. J Physiol 590:1313–1330

19. Re RN (2002) The origins of intracrine hormone action. Am J Med Sci 323:43–48

20. Re RN (2003) The intracrine hypothesis and intracellular peptide hormone action. Bioessays 25:401–409

21. Re RN, Cook JL (2006) The intracrine hypothesis: an update. Regul Pept 133:1–9

22. Re RN, Cook JL (2008) The basis of an intracrine pharmacology. J Clin Pharmacol 48:344–350

23. Re RN, Cook JL (2009) Senescence, apoptosis, and stem cell biology: the rationale for an expanded view of intracrine action. Am J Physiol Heart Circ Physiol 297:H893–H901

24. Re RN, Cook JL (2010) The mitochondrial component of intracrine action. Am J Physiol Heart Circ Physiol 299:H577–H583

25. Re RN (2011) Lysosomal action of intracrine angiotensin II. Focus on "Intracellular angiotensin II activates rat myometrium". Am J Physiol Cell Physiol 301:C553–C554

26. Kumar R, Thomas CM, Yong QC et al (2012) The intracrine renin-angiotensin system. Clin Sci (Lond) 123:273–284

27. Re RN (2013) Could intracrine biology play a role in the pathogenesis of transmissible spongiform encephalopathies, Alzheimer's Disease, and other neurodegenerative diseases? Am J Med Sci 347:312–320

Chapter 2

Single-Cell Microinjection Coupled to Confocal Microscopy to Characterize Nuclear Membrane Receptors in Freshly Isolated Cardiomyocytes

Clémence Merlen and Jonathan Ledoux

Abstract

Lipid bilayers, such as the plasma membrane and nuclear envelope, serve as effective cellular barriers to ions and macromolecules, thus allowing regulated access to subcellular compartments including the cytoplasm and nucleus, respectively. Of course, these barriers are semipermeable and a wide variety of proteins including transporters, ion exchangers, pumps, and ion channels are required to permit access as well as establish and maintain molecular and ionic gradients across membranes. However, some experimental designs, such as specifically targeting intracellular receptors, require the administration of membrane-impermeable molecules directly into live cells. The microinjection technique described in this chapter is an efficient, technically simple, and reliable approach that can be used to introduce macromolecules into intracellular compartments while maintaining the integrity of the plasma membrane itself.

Key words Microinjection, Nuclear receptor, Confocal microscopy

1 Introduction

First reported in the early twentieth century [1], microinjection has been more commonly used in the last few decades to introduce macromolecules and cell-impermeable molecules directly into the cytoplasm or nucleus. For example, this approach has been used to inject genetically manipulated embryonic stem cells into enucleated blastocysts [2]. A wide variety of molecules, ranging from proteins to DNA constructs, can therefore be introduced in cells using this versatile technique. In addition to being exceedingly efficient, its reliability allows the experimenter to inject a defined quantity of material and then quantify experimental outputs.

Investigation of intracellular receptors such as the inositol 1,4,5-phosphate receptor (IP_3R) can be carried out by cytoplasmic application of IP_3 or photosensitive caged IP_3, which can release IP_3 with a high spatial and temporal resolution and subsequently

Bruce G. Allen and Terence E. Hébert (eds.), *Nuclear G-Protein Coupled Receptors*, Methods in Molecular Biology, vol. 1234, DOI 10.1007/978-1-4939-1755-6_2, © Springer Science+Business Media New York 2015

activate IP$_3$R [3]. Similarly, identification and characterization of nuclear receptors in intact cells can be facilitated by microinjecting the ligand. Indeed, discrimination between internalized ligand-receptor complexes and the nuclear receptor can be achieved with microinjection of the ligand, as plasmalemmal receptors will not have been exposed to the ligand. For example, nuclear AngII has been investigated using this approach [4].

This chapter describes the microinjection of fluorescently labeled endothelin-3 (ET-3) into freshly isolated adult rat cardiomyocytes to allow the study of intracellular pools of the ET receptor.

2   Materials

Microinjection is performed using a Nikon Eclipse FN1 upright microscope as part of an Andor Revolution confocal system. Equivalent systems are available from other commercial sources. The use of two lasers (488 and 561 nm) and the appropriate emission filters permits concomitant injection of two different fluorophores. Images are acquired with an iXon EMCCD camera with a 60× water-dipping objective (N.A. 1.0) (*see* **Note 1**).

Air-pressure microinjection unit such as a FemtoJet system coupled to an InjectMan micromanipulator (Eppendorf) (*see* **Notes 2 and 3**).

Femtotip (Eppendorf) microinjection pipettes to access the cytosol and inject the molecule of interest (*see* **Note 4**). The microfilament in the pipette ensures that the solution reaches the tip of the microinjection pipette.

Microloader pipette tips to load the microinjection pipette with the desired solution.

All solution should be made at room temperature and prepared using fresh, type 1 deionized water (dH$_2$O).

Microinjection solution: 20 mM HEPES, 3 mM KCl, and 2 mM NaCl. The solution must be prepared fresh and, following pH adjustment (pH 7.2; with KOH), be filtered to 0.22 μm (*see* **Note 5**).

Extracellular solution (KB): 10 mM HEPES, 20 mM KCl, 10 mM KH$_2$PO$_4$, 70 mM K$^+$-glutamate, 100 mM K-aspartate, 2 mM MgSO$_4$, 5 mM creatine, 1 mM MgCl$_2$, 25 mM glucose, 10 mM β-hydroxybutyric acid, 20 mM taurine, 0.5 mM EGTA, and 0.1 % albumin. Adjust pH to 7.25 with KOH. KB solution can be stored at 4 °C for up to 7 days.

Laminin-coated cover slips: Prior to coating the cover slips, dilute laminin stock solution (1 mg/mL) to a concentration of 15 μg/mL using dH$_2$O. Apply and spread 100 μL of diluted laminin onto the cover slip. After incubating for 20 min at room temperature, remove the excess laminin solution. Cover slips must be completely dry before plating cardiomyocytes.

Rhodamine ET-3: Dilute rhodamine ET-3 in the microinjection solution to achieve a final concentration of 2 nM. Keep on ice.

Sodium fluorescein solution: Dilute sodium fluorescein to a concentration of 0.2 mM in microinjection solution.

Syto 11: Fluorescent nucleic acid stain that is used to label the nucleus.

3 Methods

3.1 Cardiomyocyte Isolation

Freshly isolated adult rat ventricular cardiomyocytes were prepared as described previously [5, 6]. Following enzymatic dissociation of ventricular cardiomyocytes, centrifuge cells for 1 min at $45 \times g$ at room temperature. Remove isolation buffer with gentle aspiration and replace with 10 mL KB buffer. Repeat this wash step three times to remove all traces of collagenase from the cell suspension. This step increases longevity and viability of the harvested cells (*see* **Note 6**). Resuspend cardiomyocytes in KB, plate 300 μL of cells on a laminin-coated cover slip, and incubate for 1 h at 37 °C in a humidified chamber. It is important to aim for a final confluence of ≈50 %. Cells must be used within 6 h of isolation.

3.2 Microinjection

When filling the Femtotip microinjection pipette, select the "capillary exchange" mode in the menu of the FemtoJet system, and then connect the injection tube. Using a Microloader, insert 1–5 μL of microinjection solution into the microinjection pipette (*see* **Note 7**). Then, place the microinjection pipette on the pipette holder on the InjectMan, secure the pipette, and press Menu to exit from "capillary exchange" mode on the FemtoJet.

Using rhodamine-conjugated ET-3 allows the use of ET-3 as a fluorescent marker for intracellular ET_B receptors. Combined injection of fluorescein and rhodamine-ET-3 enables the operator to assess the efficiency of microinjection (*see* **Notes 8** and **9**).

Prewarm the solutions and the experimental chamber. Experiments are performed at 37 °C.

Using forceps, carefully transfer a cardiomyocyte-loaded cover slip in the recording chamber (*see* **Note 10**).

Visually identify a healthy, quiescent cardiomyocyte (*see* **Note 11**) in the chamber using bright-field illumination (Fig. 1). Position the microinjection pipette beside (not above) the cell. Adjust the focal plane of the microscope to be at approximately half of the height (z-axis) of the cell. Now, move the microinjection pipette until the tip is in focus at this focal plane. The microinjection pipette is now aligned with the midpoint of the cell (on the z-axis). Set this z value as the "z limit" on the InjectMan with the option "inject at z limit." This will be the height at which the microinjection will be initiated by the system. Setting the z limit too high will result in a missed injection (release of the solution above the cell).

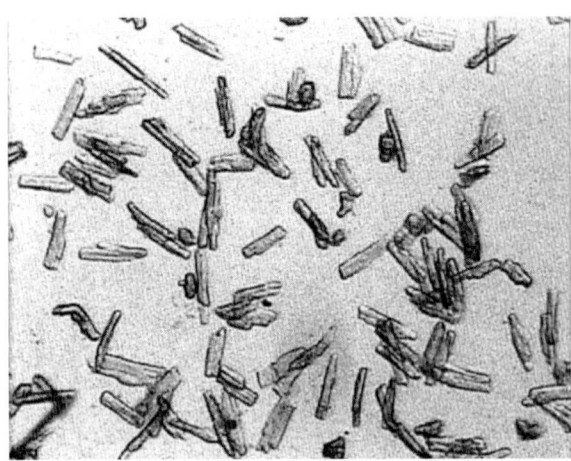

Fig. 1 Freshly isolated rat ventricular cardiomyocytes. Typical field of view of freshly isolated ventricular cardiomyocytes seeded on a laminin-coated cover slip and viewed using bright-field illumination. Note the distinctive *rectangular* shape of the cells

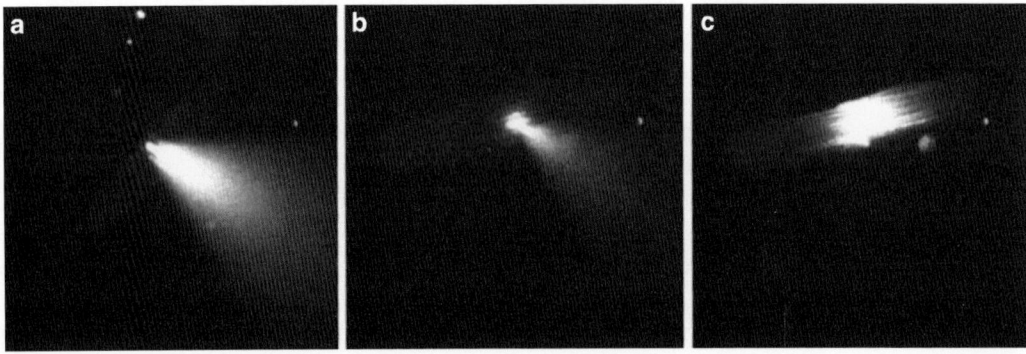

Fig. 2 Microinjection of rat ventricular cardiomyocytes with fluorescein and rhodamine-ET-3. Representative images selected from a typical microinjection experiment of a rat ventricular cardiomyocyte. Confocal images of a cardiomyocyte taken prior to (**a**), during (**b**), and 45 s following (**c**) microinjection

In contrast, using a z limit that is too low might result in the micropipette coming in contact with the bottom of the chamber and breaking (*see* **Note 12**).

Raise the focal plane of the microscope until the top of the cardiomyocyte is in focus. Now, raise the microinjection pipette until the tip is slightly above the focal plane. Finally, maneuver the microinjection pipette until the tip is above the cell at the exact *x–y* value where the microinjection will be done (*see* **Note 13**). At this point, begin recording confocal images. Pre-injection images will serve as "control" or "basal" values. Finally, initiate the microinjection by pressing on the joystick button (InjectMan) while continuously recording images to track injected fluorophores within the cell (Fig. 2).

The previous steps can be repeated on the same slide as long as healthy cardiomyocytes are available in the chamber given the microinjection pipette is not clogged.

For each cell, at the end of the experiment, add Syto 11 to the chamber at a final concentration of 1 µM to label the nucleus. This can be used to compare, for example, ET-3 staining to Syto 11 and show the localization of endothelin receptors relative to the nucleus (*see* **Note 14**).

---

4   Notes

1. Although we use an upright microscope, microinjection can also be carried out using an inverted microscope. In such cases, microinjection is easier since the objective lens of the microscope does not interfere with the microinjection system. Moreover, on an inverted microscope, the microinjection manipulator is usually attached to the body of the microscope and will stay in position with the objective while the chamber is moved around to find the best cells to inject. Alternatively, the platform where the micromanipulator of an upright scope is sitting is the same as that holding the chamber. Therefore, the microinjection pipette will move with the chamber when searching for a cell and may move out of view when the chamber is repositioned. However, an upright microscope is very useful when working with multicellular preparations and tissues, such as brain slices and whole arteries, where the cells to be imaged are located on the surface of the tissue.

2. The use of two micromanipulators, one to microinject and the other one to hold the cell, is not required in the described protocol. The use of laminin (Cell-Tak is a potential alternative) on the cover slip allows cardiomyocytes to attach to the cover slip and remain immobilized during microinjection and image acquisition. However, the use of laminin-coated plates does not prevent spontaneous contraction of cardiomyocytes.

3. The angle of the microinjection pipette is to be set between 30° and 45°. Although an angle of 45° is often preferred, with an upright microscope a different angle might be required due to the objective lens impeding access of the micropipette approaching the targeted cell.

4. In order to avoid the high cost of buying microinjection pipettes, custom-made pipettes can be generated from borosilicate glass capillaries using a regular pipette puller. However, significant optimization is required to find the best settings for pipettes to be used for microinjection. Depending on the pipette puller used, several parameters

might need to be optimized, including heating temperature, time, pulling force, and sequence of the protocol.

5. Filtration of the microinjection solution is very important to ensure that the microinjection pipette does not become clogged with debris during experiments.

6. It is important to carefully remove any phenol from the isolation buffer, as phenol interferes with fluorescence and will alter the fluorescence recordings.

7. When filling the microinjection pipette, bring the Microloader tip close to the tip of the microinjection pipette and carefully release the contents of the Microloader. Then gently flick the microinjection pipette with your finger to bring any air bubbles to the top of the microinjection solution. Air bubbles can impede injection and must be removed from the microinjection pipette. However, care must be taken here as the microinjection tips are extremely fragile.

8. Briefly centrifuge the rhodamine-ET-3 stock solution in a microcentrifuge just before diluting it in microinjection solution to avoid blocking the microinjection pipette. Co-injecting rhodamine-ET-3 with fluorescein allows the operator to assess (1) loading of the microinjection pipette, (2) the quality of the microinjection (e.g., fluorescein will reveal the entry and diffusion of the microinjection buffer through the cell as well as any leakage at the site of injection), (3) the compensation pressure (*see* below), and (4) if the pipette is blocked.

9. In the current protocol, fluorescein was chosen to be co-injected with rhodamine-ET-3 as it permits concomitant recording. Alternatively, if the ligand of interest is conjugated with a dye that requires excitation in the 488 nm range, a dye such as Texas Red can be co-injected, depending on the laser/filter availability of the imaging system. However, autofluorescence can be a problem when cardiomyocytes are excited by a 488 nm laser, especially when using low-quantum-yield dyes or studying low-density target proteins.

10. Compensation pressure is an important parameter that must be set prior to performing actual experiments. The compensation pressure is defined as the pressure applied by the system to compensate for the differential force of the capillary suction and hydrostatic pressure. The compensation pressure can be determined with a fluorescein-loaded microinjection pipette. Using the confocal system, while looking at the tip of the microinjection pipette, adjust the compensation pressure on the FemtoJet system to ensure that there is no leak of fluorescent solution. Then, injection pressure and time should be set, again under control conditions with only fluorescein in the

pipette. Once the different pressure and duration parameters defined, the injection volume can be determined. Load a microinjection pipette with 1 µL of solution and repeatedly activate the injection until the microinjection pipette is empty. The volume of each injection can then be assessed by dividing 1 µL by the number of injections required to empty the pipette.

11. Quiescent (non-beating) cardiomyocytes are considered healthy cells as their Ca^{2+} homeostasis mechanisms remain able to prevent Ca^{2+} overload and cell death. Unhealthy cardiomyocytes develop blebs on their surface and then hyper-contract. Moreover, moving cardiomyocytes are extremely difficult to microinject without either damaging the microinjection pipette or the target cell.

12. To a neophyte, an unexpectedly delicate part in the microinjection procedure is to locate the tip of the microinjection pipette in order not to crush the tip against the bottom of the chamber. Use a lower magnification objective lens to find the pipette when it first enters the solution. The halo created by surface tension can be used to quickly localize the tip of the microinjection pipette. Once found, the tip can be lowered into the chamber while simultaneously adjusting the focus to be ahead of the tip. When the outline of the cells becomes visible, stop lowering the microinjection pipette and change the step size of the micromanipulator to the fine or extra-fine setting.

13. In experiments where the objective is to study nuclear receptors, it is important not to aim the microinjection pipette at the nucleus of the cell (during injection, the microinjection pipette may actually damage the nucleus). Hence, when positioning the microinjection pipette in the x–y plane over the cell, choose an injection site that is not too close to the nuclei (adult cardiomyocytes are usually binucleate). For example, the microinjection site can be halfway between the two nuclei. The nuclei are clearly visible when using bright-field illumination.

14. Although the nuclei in cardiomyocytes can be identified as elliptical regions devoid of fluorescence from the other injected fluorophores, the use of a cell-permeable nucleic acid stain such as Syto 11 is a preferable means of identifying nuclei. Syto 11 has an excitation/emission of 508/527 nm and can therefore be viewed with a 488 laser and filters as for FITC.

Acknowledgements

This work is supported by CFI, FRQS, and HSFC.

References

1. Barber MA (1911) A technic for the inoculation of bacteria and other substances into living cells. J Infect Dis 8:348–360

2. Hooper M, Hardy K, Handyside A et al (1987) Hprt-deficient (lesch-nyhan) mouse embryos derived from germline colonization by cultured cells. Nature 326:292–295

3. Parker I, Ivorra I (1991) Caffeine inhibits inositol trisphosphate-mediated liberation of intracellular calcium in xenopus oocytes. J Physiol 433:229–240

4. Tadevosyan A, Maguy A, Villeneuve LR et al (2010) Nuclear-delimited angiotensin receptor-mediated signaling regulates cardiomyocyte gene expression. J Biol Chem 285:22338–22349

5. Rodrigues B, Severson DL (1997) Preparation of cardiomyocytes. In: McNeill JH (ed) Biochemical techniques in the heart. CRC, New York, NY, pp 101–115

6. Chevalier D, Allen BG (2000) Two distinct forms of mapkap kinase-2 in adult cardiac ventricular myocytes. Biochemistry 39:6145–6156

Chapter 3

Design and Application of Light-Activated Probes for Cellular Signaling

David Chatenet, Steve Bourgault, and Alain Fournier

Abstract

Multiple reports have described the presence of functional G protein-coupled receptors (GPCRs) in the perinuclear/nuclear membranes of many cell types where they are able to modulate nuclear Ca^{2+} influx, transcription initiation, and gene expression. Because GPCRs represent "some of the most promising targets for drug development" a better understanding of their roles, not only at the cell membrane but also at the nuclear level, in healthy and disease states, will certainly generate new avenues for therapeutic intervention. The photo-triggered release of biologically active compounds has been regarded as one of the most effective methods for inducing an in vitro-controlled biochemical or physiological response. Here, we describe various methodologies and alternatives related to the conception of inert biologically active peptides through the incorporation of photo-triggered groups at key positions of the native peptide sequence.

Key words Caged amino acids, Caged peptides, Orthogonal cleavage, Photo-click O-acyl isopeptide, Solid-phase peptide synthesis

1 Introduction

In the last decades, several G protein-coupled receptors (GPCRs) have been observed at the nuclear membrane where they exert multiple functions ranging from nuclear Ca^{2+} influx, transcription initiation, and gene expression [1]. The existence of nuclear-localized receptors suggests that the biological responses mediated by GPCRs are not solely initiated at the cell surface but might result from the integration of extracellular and intracellular signaling pathways [2]. Accordingly, a better understanding of the specific roles played by those nuclear receptors is critical to assess their potential therapeutic value. Although caged compounds of low molecular weight, such as inositol 1,4,5-triphosphate [3] or glutamate [4], have been developed and used effectively to study the mechanisms of temporal biological phenomena, such as muscle contraction, intracellular signaling, and neurotransmission, a very

Bruce G. Allen and Terence E. Hébert (eds.), *Nuclear G-Protein Coupled Receptors*, Methods in Molecular Biology, vol. 1234, DOI 10.1007/978-1-4939-1755-6_3, © Springer Science+Business Media New York 2015

limited number of caged endogenous peptides or related analogs have been reported [5–8]. Nevertheless, the development of light-activated peptidic probes provides an elegant and efficient way to control activation of their responsive nuclear GPCRs, yielding an unparalleled degree of temporal and spatial resolution [7]. Basically, a biologically inert derivative is delivered into a cell and allowed to diffuse and equilibrate throughout the cytosol. Upon irradiation with pulsed and focused UV light, the concentration jump of biologically active substance can be brought about immediately in a spatially controlled area. Such an approach, allowing rapid (milliseconds), spatially defined release of the native derivative without compromising cell organization and integrity, may definitively help us understand the physiological roles of these recently discovered nuclear GPCRs.

Several nuclear GPCRs participate in the control of cardiovascular or nervous system homeostasis by interacting with their endogenous peptidic ligands [1, 2]. Despite the revival of light-triggered moieties, the design of light-activated probes is not trivial and typically takes advantage of known structure-activity relationship data [9, 10]. Hence, following the identification of key functional residues, a photosensitive derivative of relevant pharmacophoric amino acids is introduced into the peptide sequence, thereby rendering the agent potentially biologically silent. In some cases, inactivation of the peptide through this approach is not possible because residual activity remains too high for in vitro application. Ligand recognition and activation of GPCRs are generally driven by interactions involving specific residues/functions that are embedded within a prerequisite secondary conformation of the ligand [7]. Other strategies could thus rely on control of the secondary structure of the peptide via the introduction of photocleavable groups on the peptide backbone amide [11] or by a photo-triggered $O \rightarrow N$ intramolecular acyl migration [12]. In the next section, we describe various methodologies to prepare caged amino acids that can be introduced within a peptide sequence, as well as procedures for introduction of a caged group within a peptide amide backbone or for the synthesis of photo-click *O*-acyl isopeptide.

2 Materials

1. All chemical reagents and solvents can be obtained from specialized suppliers:
   1. S-methylisothiourea semisulfate.
   2. 6-Nitroveratryl chloroformate.
   3. 2-Nitrobenzyl bromide.
   4. 2,6-Dimethylpyridine.
   5. Di-*tert*-butyl dicarbonate.

6. Potassium *tert*-butoxide.

7. 4,5-Dimethoxy-2-nitrobenzyl bromide.

8. 4,5-Dimethoxy-2-nitrobenzyl chloroformate.

9. 2-Nitrobenzaldehyde.

10. 2-Nitrobenzylamine.

11. Dimethoxy-2-nitrobenzyl alcohol.

12. Ethanedithiol.

13. Phenol.

14. *N*,*N*-diisopropylethylamine.

15. Magnesium sulfate.

16. Tetrahydrofuran.

17. Trifluoroacetic acid.

18. Chloroform.

19. Methanol.

20. Acetic acid.

21. Acetonitrile (ACN).

22. *N*-(9-fluorenylmethoxycarbonyloxy)-succinimide.

23. Triethylamine.

24. Ethyl acetate.

25. Sodium hydroxide.

26. Hexane.

27. Dimethylformamide.

28. Benzene.

29. Ether.

30. Piperidine.

31. Diisopropylcarbodiimide.

32. *N*-methylmorpholine.

33. *N*-ethylmorpholine.

34. 4-Dimethylaminopyridine.

35. 1,3-Dicyclohexylcarbodiimide.

2. Brine: Add 40 g of NaCl in 100 mL of water.

3. All amino acid derivatives, resins, and reagents for solid-phase peptide synthesis can be purchased from specialized suppliers:

1. H-Tyr-α-*tert*-butyl ester.

2. *N*-α-Fmoc-L-lysine-OH.

3. *N*-α-Fmoc-L-ornithine-OH.

4. *N*-α-Boc-L-histidine-α-*tert*-methyl ester.

5. H-Gly-OH.

6. N-α-Fmoc-L-aspartic acid-α-*tert*-butyl ester.

7. N-α-Fmoc-L-glutamic acid-α-*tert*-butyl ester.

8. N-α-Fmoc-L-cysteine-OH.

9. N-α-Boc-L-aspartic acid-α-*tert*-butyl ester.

10. N-α-Boc-L-glutamic acid-α-*tert*-butyl ester.

11. O-(6-chlorobenzotriazol-1-yl)-N,N,N',N'-tetramethyluronium hexafluorophosphate.

12. N-hydroxybenzotriazole.

13. Wang resin.

14. Rink amide resin.

15. 2-Chlorotrityl chloride resin.

4. Reagent A: Dissolve 1 g of ninhydrin in 20 mL of ethanol. Store at room temperature in the dark.

5. Reagent B: Dissolve 40 g of phenol in 20 mL of ethanol. Store at room temperature in the dark.

6. Reagent C: Dissolve 16.5 mg of KCN in 25 mL of distilled water and dilute 1 mL of this solution with 49 mL of pyridine. Store at room temperature in the dark.

Reverse-phase high-performance liquid chromatography (RP-HPLC):

1. Solvent A: H_2O (0.06 % TFA, v/v).

2. Solvent B: Acetonitrile (ACN).

3. Analytical RP-HPLC analyses were performed on a Jupiter C_{18} (5 μm, 300 Å, 250 mm×4.6 mm) column (Phenomenex, Torrance, CA).

4. Preparative RP-HPLC was performed using a C_{18} (15 μm, 300 Å, 250 mm×21.2 mm) column (Phenomenex).

3   Methods

To prevent unwanted photocleavage, all reactions should be carried out in the dark by putting aluminum foil around the reaction vessel. The general procedure described herein employs conditions that minimize racemization of the amino acid and provides a simple, high-yield route to caged amino acids.

3.1   Preparation of Caged Fmoc-Tyr (4-ODMNB)-OH (See Note 1) [5]

1. To a solution of H-L-tyrosine-α-*tert*-butyl ester (H-Tyr-OtBu; 1 equiv.) dissolved in dioxane-H_2O (5:1, v/v) is added 1 equiv. of N,N-diisopropylethylamine (DIPEA) and di-*tert*-butyl dicarbonate (Boc$_2$O, 1.1 equiv.). The mixture is stirred at room temperature for 2 h.

2. After evaporation, the solid is extracted three times with dichloromethane (DCM)-H$_2$O. The organic layers are dried over anhydrous MgSO$_4$ and evaporated to dryness in vacuo.

3. Boc-Tyr-OtBu (1 equiv.) is then dissolved in dry tetrahydrofuran (THF) and 1.1 equiv. of potassium *tert*-butoxide is added under N$_2$ atmosphere, followed by 1.1 equiv. of 4,5-dimethoxy-2-nitrobenzyl bromide (DMNBB) also dissolved in dry THF. The solution is stirred for 16 h under N$_2$. Progress of the reaction is monitored by RP-HPLC or thin-layer chromatography (TLC) analysis (chloroform-methanol-acetic acid; 94:5:1, v/v).

4. After evaporation, the crude product is extracted three times with DCM-H$_2$O. The organic phases are dried over anhydrous MgSO$_4$ and evaporated.

5. The resulting oil (Boc-Tyr(4-ODMNB)-OtBu) is dissolved in trifluoroacetic acid (TFA)-H$_2$O (19:1, v/v), and stirred for 2 h at room temperature.

6. After completion of deprotection (monitored by HPLC), the solvent is evaporated in vacuo. The resulting oil is resuspended in acetonitrile (ACN)-H$_2$O (2:3, v/v) and purified by RP-HPLC (C$_{18}$ column, 15 μm, 300 Å, 250×21.2 mm) using a linear gradient of 40–70 % solvent B. The pure product (H-Tyr(4-ODMNB)-OH) is obtained after evaporation of the appropriate fractions.

7. The product H-Tyr(4-ODMNB)-OH (1 equiv.) is dissolved in ACN-H$_2$O-triethylamine (TEA) (5:2:0.2, v/v/v) and 1 equiv. of N-(9-fluorenylmethoxycarbonyloxy)-succinimide (Fmoc-OSu) dissolved in ACN-H$_2$O (5:2, v/v) is added. The mixture is stirred for 1 h at room temperature. The pH of the reaction mixture is maintained at 8.5 by addition of TEA.

8. The solvent is evaporated and the resulting oil is extracted three times with ethyl acetate (EtOAc)-H$_2$O. The organic phases are isolated, dried over anhydrous MgSO$_4$, and evaporated. If necessary, the Fmoc-protected amino acid derivative is purified using RP-HPLC to obtain pure Fmoc-Tyr(4-ODMNB)-OH.

3.2 Preparation of Fmoc-Lys(DMNB)-OH [5]

1. To a solution of N-α-Fmoc-L-lysine-OH (1 equiv.) dissolved in THF is added 1 equiv. of 4,5-dimethoxy-2-nitrobenzyl chloroformate (DMNBC) previously dissolved in THF-DCM (5:1, v/v), containing 2 equiv. of DIPEA. The reaction mixture is stirred at room temperature for 2 h (*see* **Note 2**).

2. After evaporation in vacuo, the crude product (Fmoc-Lys (DMNB)-OH) is extracted three times with EtOAc-H$_2$O. The organic phases are collected, dried over anhydrous MgSO$_4$, and evaporated. If necessary, the resulting Fmoc-protected Lys derivative is purified by RP-HPLC.

3.3 Preparation of Fmoc-Arg (DMNB)- OH [13]

The initial step is the preparation of a guanylating agent formed by the condensation of S-methylisothiourea semisulfate and 6-nitroveratryl chloroformate.

1. To a suspension of S-methylisothiourea semisulfate (1 equiv.) in DCM is added a solution of sodium hydroxide (NaOH, 4 N, 0.5 equiv.) and the mixture is cooled to 0 °C.

2. Then, 6-nitroveratryl chloroformate (0.5 equiv.) is dissolved in a minimum volume of DCM and added dropwise to the above suspension (*see* **Note 3**) maintained at 0 °C. After the addition, the reaction is stirred overnight at room temperature.

3. The solution is extracted three times with DCM and all organic layers are combined, and washed three times with 0.1 N HCl, water, and brine, before being dried over anhydrous $MgSO_4$.

4. The solution is filtered and the purity of the product is evaluated by TLC (hexane-EtOAc, 70:30) or RP-HPLC.

5. The integrity of the derivative can be assessed by nuclear magnetic resonance (NMR). Spectral data of this derivative are found in Wood et al. [13].

6. To a solution of Fmoc-Orn-OH (1 equiv.), dissolved in dimethylformamide (DMF), are added 10 equiv. of the guanylating agent and 20 equiv. of 2,6-dimethylpyridine. The reaction is stirred and the progress monitored by RP-HPLC.

7. Upon completion, EtOAc is added and the organic layer is washed with H_2O, dried with anhydrous MgSO4, and concentrated in vacuo. If necessary, the resulting residue is purified by flash chromatography (SiO_2) to obtain Fmoc-Arg(DMNB)-OH.

3.4 Synthesis of Fmoc-His(ONB)- OH [14]

As a first step, a Boc-His-OMe silver salt must be prepared.

1. A solution of silver nitrate (1 equiv.) in dimethylsulfoxide (DMSO) is added to a solution of Boc-His-OMe (1 equiv.) in absolute ethanol and to this mixture is added dropwise an ethanolic sodium hydroxide solution (0.4 N, 1 equiv.) over a period of 2 h while the mixture is stirred vigorously. The off-white solid is isolated by suction on filter paper and thoroughly washed with absolute ethanol. The precipitate is suspended in water, stirred vigorously for several minutes, filtered, and washed again with ethanol. This process was repeated in acetone and the solid was dried in a vacuum desiccator (*see* **Note 4**).

2. A solution of this solid (1 equiv.) in benzene is refluxed for 4 h with 2-nitrobenzyl bromide (ONB, 1 equiv.).

3. The silver salt is then removed by filtration and the crude methyl ester product is saponified by treatment with 2 equiv. of NaOH (1 N) in MeOH-DMF (3:1) for 2 h.

4. Upon completion, the solution is diluted with water, and then extracted with EtOAc.

5. The amino acid is then dissolved in TFA-H$_2$O (19/1, v/v), and stirred for 2 h at room temperature. After the α-amine deprotection (monitored by HPLC), the solvent is evaporated in vacuo.

   As a final step, an Fmoc-protecting group is added on the free α-amine of the His derivative.

6. Fmoc-OSu (0.95 equiv.) is added to a solution of H-His(ONB)-OH (1 equiv.) and Na$_2$CO$_3$ (1 equiv.) in H$_2$O-acetone (1:1) over a period of 1 h while the pH is kept between 9 and 10 by the addition of 1 M Na$_2$CO$_3$ solution. Stirring is continued overnight and progression of the reaction is monitored by RP-HPLC.

7. Upon completion, EtOAc is added and the mixture is acidified with 6 M HCl. The organic layer is washed with H$_2$O, dried over anhydrous MgSO$_4$, and concentrated in vacuo. If necessary, the resulting residue is purified by flash chromatography (SiO$_2$).

3.5   Preparation of Fmoc-N-α-(2-Nitrobenzyl)-Gly-OH [11]

1. To a solution of 2-nitrobenzaldehyde (NBA) (1.1 equiv.) in methanol is added 1 equiv. of glycine dissolved in methanol and an equivalent volume of 2 N NaOH.

2. After stirring for 1 h at room temperature, NaBH$_4$ (3 equiv.) is added to the solution at 0 °C and the reaction mixture is stirred for 2 h at 0 °C.

3. The solution is then evaporated and washed with diethyl ether. Water is added and the pH of the aqueous phase is adjusted to 5 with HCl, before being washed with diethyl ether. The aqueous phases are pooled and evaporated to give crude N-α-nitrobenzyl-Gly-OH.

4. To a solution of N-α-nitrobenzyl-Gly-OH (1 equiv.) in 10 % NaHCO$_3$-H$_2$O-acetone (30:70:80, v/v) is added 1.5 equiv. of fluorenylmethyloxycarbonyl chloride (Fmoc-Cl) dissolved in acetone. The reaction mixture is stirred for 16 h at room temperature.

5. Upon completion, the acetone is evaporated and the resulting suspension is washed three times with diethyl ether. The pH of the aqueous phase is adjusted to 4 with HCl and EtOAc is added. The organic phases are washed with water and dried over anhydrous MgSO$_4$.

6. Fmoc-N-α-(2-nitrobenzyl)-Gly-OH is obtained as a white solid upon precipitation with hexane.

3.6 Preparation of Fmoc-Asp(ODMNB)-OH and Fmoc-Glu (ODMNB)-OH [6]

1. To a solution of *N*-α-Fmoc-L-aspartic acid-α-*tert*-butyl ester (Fmoc-Asp-OtBu, 1 equiv.) or *N*-α-Fmoc-L-glutamic acid-α-tert-butyl ester (Fmoc-Glu-OtBu, 1 equiv.) dissolved in DCM are added 1 equiv. of dimethoxy-2-nitrobenzyl alcohol (DMNBA) dissolved in DCM and 0.05 equiv. of 4-dimethyl-aminopyridine (DMAP). Then, 1,3-dicyclohexylcarbodiimide (DCC; 1 equiv.) dissolved in DCM is added dropwise.

2. After 2 h of agitation at room temperature, the mixture is filtered and washed three times with aqueous 2.5 % NaHCO$_3$. The organic phases are combined and evaporated to dryness in vacuo.

3. The resulting oil is resuspended in TFA-H$_2$O (19:1) and stirred at room temperature for 90 min to remove the α-carboxyl-protecting group.

4. After solvent evaporation, the crude product is extracted with DCM-H$_2$O and the organic phases are collected, dried over anhydrous MgSO$_4$, and evaporated. If necessary, the resulting Fmoc-protected Asp (or Glu) residue is purified by RP-HPLC, to obtain the pure product.

3.7 Preparation of Fmoc-Cys (DMNB)-OH (Adapted from Pan and Bayley [15])

1. To a solution of *N*-α-Fmoc-L-cysteine-OH (Fmoc-Cys-OH) (1 equiv.) dissolved in DCM is added 1.1 equiv. of 4,5-dimethoxy-2-nitrobenzyl bromide (DMNBB) previously dissolved in DCM. The reaction mixture is stirred for 1 h at room temperature.

2. Upon completion, the reaction mixture is washed three times with aqueous 2.5 % NaHCO$_3$. The aqueous layers are then combined, acidified with 6 N HCl, and extracted with DCM three times. The organic phases are collected, dried over anhydrous MgSO$_4$, and evaporated.

3. If necessary, the resulting Fmoc-L-Cys(DMNB)-OH is purified by RP-HPLC.

3.8 Preparation of Fmoc-Gln(2-NB)-OH and Fmoc-Asn (2-NB)-OH [16]

The first step involves the formation of an amide bond between the amine of the photolabile group, i.e., 2-nitrobenzylamine (2-NB), and the unprotected carboxylic acid group on the N-Boc- and OtBu-protected amino acid derivatives.

1. To a solution of 2-nitro-benzylamine (0.9 equiv.) in DCM cooled at 0 °C are added Boc-Glu-OtBu or Boc-Asp-OtBu (1 equiv.), 1-(3-dimethylaminopropyl)-3-ethylcarbodiimide (EDCI, 0.9 equiv.), 4-(dimethylamino)pyridine (DMAP, 0.1 equiv.), and *N*-hydroxybenzotriazole (HOBt, 0.1 equiv.).

2. Then, *N*-ethylmorpholine (NME, 1.05 equiv.) is added dropwise over 30 min. During the addition, the temperature is controlled and kept at 0 °C (*see* **Note 3**). Progression of the reaction is monitored by RP-HPLC.

3. Upon completion, the solvent is evaporated, and then chilled water is added. The content is extracted three times with EtOAc. The combined organic layers are successively washed three times with a saturated $NaHCO_3$ solution and a 10 % citric acid solution, and then once with water and brine. NMR spectral characterization of Boc-Gln(2-NB)-OtBu and Boc-Asn(2-NB)-OtBu can be found elsewhere [16].

4. The organic solvent is concentrated and the deprotection of the N- and C-terminal protecting groups is achieved at 0 °C by adding TFA-H_2O (30:70, v/v). The solution is allowed to stir for 20 h at room temperature. Ether is then slowly added by portion and the precipitated solid is filtered, washed with DCM, and dried [16]. Confirmation of the purity is assessed by RP-HPLC.

   As a final step, an Fmoc-protecting group is added on the free amino terminus of the amino acid derivatives.

5. Fmoc-OSu (0.95 equiv.) is added to a solution of H-Gln (2-NB)-OH (1 equiv.) and Na_2CO_3 (1 equiv.) in H_2O-acetone (1:1) over a period of 1 h while the pH is kept between 9 and 10 by the addition of 1 M Na_2CO_3 solution. Stirring is continued overnight and progression of the reaction is monitored by RP-HPLC.

6. Upon completion, EtOAc is added and the mixture is acidified with 6 M HCl. The organic layer is washed with H_2O, dried over anhydrous MgSO4, and concentrated in vacuo. If necessary, the resulting residue is purified by flash chromatography (SiO_2).

3.9 Introduction of a Caged Amino Acid Within a Peptide Sequence

Usually, peptides are produced either with a carboxylic or a carboxamide C-terminal function. Depending on the desired C-terminal group, i.e., carboxyl or amide, Wang resin preloaded with the last amino acid of the peptide sequence or Rink amide resin is generally used. In the latter case, the last amino acid is directly coupled to the Rink amide resin through conventional method.

3.9.1 Resin

3.9.2 Amino Acid Coupling

The Fmoc-amino acid (3 equiv. based on the resin substitution) and O-(6-chlorobenzotriazol-1-yl)-N,N,N',N'-tetramethyluronium hexafluorophosphate (HCTU; 3 equiv.) are placed in a beaker. A minimal volume of DMF and 6 equiv. of DIEA are added and the mixture is allowed to react for 2–3 min. This pre-activated amino acid solution is poured onto the resin already swollen in DMF and an additional volume of DMF is introduced in the reactor to allow a proper agitation. The mixture is allowed to react for about 1 h. The solution is then removed and the resin is washed twice, successively with DMF, methanol, and DCM. Completion of the reaction is monitored by the colorimetric ninhydrin test (Kaiser test—*see* Subheadings 2 and 3.9.4 below). This is a general procedure; other coupling reagents can be used. Also, some derivatives, such as Fmoc-Asp(ODMNB)-OH and Fmoc-Glu(ODMNB)-OH, require particular strategies for their use (*see* **Note 5**).

3.9.3 Deprotection of the N-α-Amine Function (Fmoc Removal)

Piperidine in DMF (10 mL, 20 %, v/v) is added to the reactor and the resin suspension is stirred vigorously for 10 min. The solution is removed and the treatment is repeated for another 10 min. After filtration of the piperidine-DMF mixture, the resin is washed twice, successively with DMF, methanol, and DCM.

3.9.4 Monitoring Acylation and Deprotection Reaction

The Kaiser test is a qualitative test for amines (*see* **Note 6**). This test is carried out by adding four drops of an ethanolic solution of ninhydrin (Kaiser test reagent A), two drops of an ethanolic solution of phenol (Kaiser test reagent B), and two drops of a potassium cyanide solution in pyridine (Kaiser test reagent C) to the test sample (usually 2–3 mg of peptidyl-resin contained in a small glass tube). The tube is heated at 100 °C for 5 min. A blue coloration of the beads or the solution is a positive result indicating that the coupling reaction is incomplete (*see* **Notes 6** and **7**).

3.9.5 Cleavage of Peptides from the Resins

Cleavage is performed with trifluoroacetic acid, in which are added a few scavengers, such as ethanedithiol (EDT), phenol, and water, to ensure the integrity of the final product. The scavenger mixture is dependent on the sequence of the peptide to be cleaved. To select the appropriate cleavage mixture, the following decisional scheme might be used (Fig. 1) (adapted from Novabiochem Fmoc resin cleavage and deprotection; www.novabiochem.com). It is generally admitted that 1 g of peptidyl-resin requires 15–20 mL of cleavage mixture and that the cleavage is performed for 90 min. Upon completion, the mixture is filtered and the resin washed twice with small volumes of TFA. TFA is then evaporated under reduced pressure until a thick oil is left. Diethyl ether is added (at least 50 mL per gram of peptidyl-resin) to precipitate the peptide and extract the scavengers. The peptide is left aside to settle and diethyl ether is removed by filtration. This operation is repeated twice and the crude peptide is air-dried. The material is purified using RP-HPLC, with ACN and aqueous TFA (0.06–0.1 %) as a binary solvent mixture. An ACN gradient is used to elute the peptide from the column. Fractions containing the desired peptide, identified by mass spectrometry, and with a satisfactory purity (>95 %), are pooled. ACN is partly evaporated and the remaining aqueous solution is frozen and lyophilized.

3.10 Introduction of a Caging Group at the Peptide C-Terminus [6]

1. The peptide is synthesized using the procedure described above (Subheading 3.8) with the exception that a highly acid-sensitive 2-chlorotrityl chloride resin is used.

2. After the coupling of the N-terminal residue, no deprotection of the amino-terminal protecting group (Fmoc) is performed.

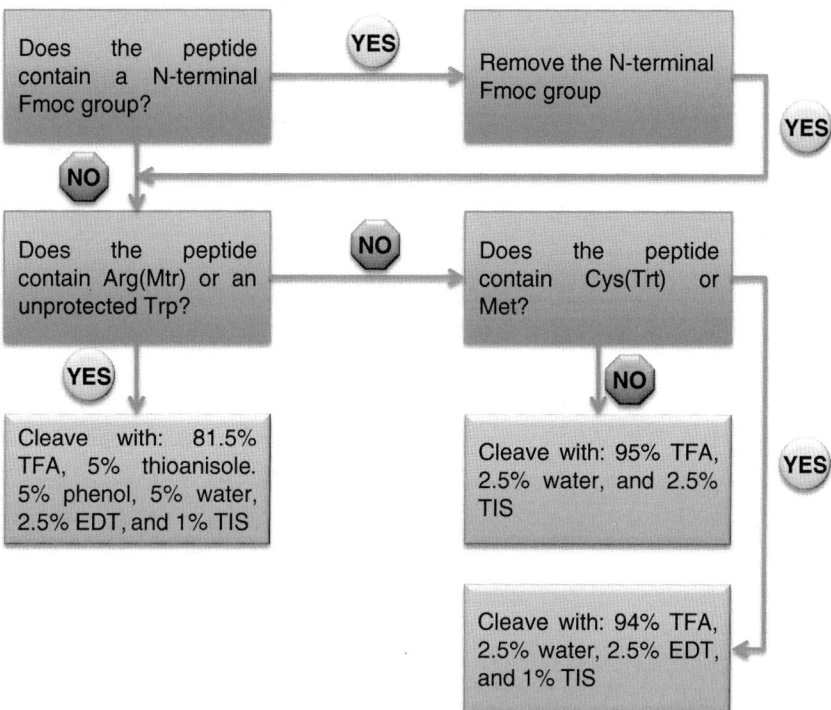

Fig. 1 Flow chart for selecting cleavage cocktail for Fmoc solid-phase peptide synthesis

3. Cleavage of the peptide from the resin is achieved with a mixture of DCM-TFA-ethanedithiol (98:1:1) [17]. The cleavage is performed over 60 min and upon completion, the resin is filtered and washed twice with DCM, the solvent is evaporated, and the protected peptide is extracted with H_2O/ DCM. The organic phases are dried over anhydrous $MgSO_4$ and evaporated. Thus, a fully protected peptide with a free carboxy-terminal function is obtained.

4. To a solution of protected peptide in DCM ($\cong$1 g/100 mL) are added 20 equiv. of dimethoxy-2-nitrobenzyl alcohol (DMNBA) dissolved in DCM and 0.2 equiv. of DMAP. Then, DCC (10 equiv.) dissolved in DCM is added dropwise and the reaction mixture is stirred overnight at room temperature.

5. After filtration of the urea by-product, the solvent is evaporated and the resulting compound is extracted with DCM-H_2O. The organic phases are dried over anhydrous $MgSO_4$ and evaporated.

6. The protected and cage-functionalized peptide is dissolved in DCM-piperidine (80:20) to remove the N-terminal Fmoc-protecting group. The mixture is allowed to react for 20 min.

Then, the reaction mixture is extracted with 1 M HCl-DCM. The organic phases are washed once with brine, dried over anhydrous MgSO₄, and evaporated in vacuo.

7. The side chain-protecting groups are finally removed using the TFA cleavage procedure described above (Subheading 3.8, **step 5**).

3.11 Synthesis of a Peptide with a Caging Group on the Peptide Backbone [11]

1. Peptide is assembled on solid phase, as described above, and Fmoc-*N*-α-(2-nitrobenzyl)-Gly-OH (*see* Subheading 3.5) is incorporated into the peptide sequence using a standard coupling strategy.

2. After the removal with piperidine of the Fmoc moiety from the *N*-α-amine of Gly (*see* Subheading 3.9, **step 3**), the next amino acid (10 equiv. vs. peptidyl-resin) is introduced in the peptide chain after its activation with 10 equiv. of diisopropylcarbodiimide (DIPCDI) and N-methylmorpholine (1.5 equiv.) in DMF. Due to the presence of a secondary amine, the qualitative Kaiser test cannot be used to follow the progression of the coupling (*see* **Note 6**).

3. The peptide synthesis is resumed as described before, and after the cleavage from the resin, the peptide is purified by RP-HPLC.

3.12 Synthesis of a Photo-Triggered Click O-Acyl Isopeptide [12]

To prepare a photo-triggered click peptide, a photocleavable group, such as 4,5-dimethoxy-2-nitrobenzyl group (DMNB) [12], needs to be attached to the α-amino group of a Ser residue. This is initially accomplished by (1) preparation of the serine derivative, (2) coupling of the serine derivative to the peptidyl-resin, (3) esterification of the β-hydroxy group of Ser with the next amino acid residue, and (4) conventional peptide synthesis.

1. To a solution of H-Ser-OH (1 equiv.) dissolved in H₂O-diethyl ether are added NaHCO₃ (2.6 equiv.) and 4,5-dimethoxy-2-nitrobenzyl chloroformate (DMNBC; 0.5 equiv.). The reaction mixture is stirred at 0 °C. After 2 h, the same quantity of NaHCO₃ and 4,5-dimethoxy-2-nitrobenzyl chloroformate is added, and the reaction mixture is stirred at room temperature for 18 h. After evaporation, the residual aqueous phase is acidified to pH 2 with HCl and extracted with EtOAc. The organic phases are dried over anhydrous MgSO₄ and evaporated to give the pure final product, *N*-6-nitroveratryloxycarbonyl-L-serine (Nvoc-Ser-OH).

2. The resulting Nvoc-Ser-OH is introduced into the peptide sequence using the strategy described in Subheading 3.9.2.

3. The Fmoc-protected amino acid next to the N-α-caged serine is coupled to the β-hydroxyl group of serine using 1 equiv. of N,N'-diisopropylcarbodiimide (DIPCDI) and DMAP in DCM for 2 h. This reaction is repeated at least four times to ensure completion.

4. The peptide synthesis is then resumed as described above and after cleavage from the resin, the peptide is purified by RP-HPLC.

4   Notes

1. This synthetic route can also be applied to prepare Fmoc-Ser(DMNB)-OH and Fmoc-Thr(DMNB)-OH.

2. Progress of the reaction can be monitored with the qualitative Kaiser test (Subheading 3.9.4).

3. It is extremely important to control the temperature since the reaction is highly exothermic and the derivative flammable.

4. More physicochemical details can be found in Kalbag and Roeske [14].

5. The use of the Fmoc-Asp(ODMNB)-OH and Fmoc-Glu(ODMNB)-OH derivatives requires particular strategies. As a matter of fact, formation of an aminosuccinyl derivative, through cyclization of the Asp(ODMNB) residue, and the formation of a pyrrolidone ring from the Glu(ODMNB) residue are highly favored by the electronic properties of the photocleavable group. Nonetheless, derivatives with caged aspartic acid could be obtained by carrying out the synthesis by incorporating next to the Asp(ODMNB) residue, a N-hydroxymethoxybenzyl (Hmb)-derivatized amino acid that would prevent, by masking the backbone amide bond, the Asp ring closure during the synthesis steps. Similarly, one possible approach to suppress the pyrrolidone formation could be the use of a N-Hmb-derivatized-Glu(ODMNB) residue.

6. This test is not reliable for the detection of proline and other N-substituted amino acids. In such a case, a double coupling is generally carried out before performing the deprotection. If the N-terminal amino acid is proline, pipecolic acid, or tetrahydroisoquinoline-3-carboxylic acid, another test, such as the isatin or the chloranil tests, is used.

7. It is important to note that certain amino acid residues will give unusual colorations ranging from red to blue (Asn, Cys, Ser, and Thr).

References

1. Chatenet D, Nguyen TT, Letourneau M, Fournier A (2012) Update on the urotensinergic system: new trends in receptor localization, activation, and drug design. Front Endocrinol (Lausanne) 3:1

2. Tadevosyan A, Vaniotis G, Allen BG et al (2012) G protein-coupled receptor signalling in the cardiac nuclear membrane: evidence and possible roles in physiological and pathophysiological function. J Physiol 590:1313–1330

3. Gjerstad J, Valen EC, Trotier D, Doving K (2003) Photolysis of caged inositol 1,4,5-trisphosphate induces action potentials in frog vomeronasal microvillar receptor neurones. Neuroscience 119:193–200

4. Callaway EM, Katz LC (1993) Photostimulation using caged glutamate reveals functional circuitry in living brain slices. Proc Natl Acad Sci U S A 90:7661–7665

5. Bourgault S, Letourneau M, Fournier A (2005) Development and pharmacological characterization of "caged" urotensin II analogs. Peptides 26:1475–1480

6. Bourgault S, Letourneau M, Fournier A (2007) Development of photolabile caged analogs of endothelin-1. Peptides 28:1074–1082

7. Lee HM, Larson DR, Lawrence DS (2009) Illuminating the chemistry of life: design, synthesis, and applications of "caged" and related photoresponsive compounds. ACS Chem Biol 4:409–427

8. Shigeri Y, Tatsu Y, Yumoto N (2001) Synthesis and application of caged peptides and proteins. Pharmacol Ther 91:85–92

9. Hruby VJ (2002) Designing peptide receptor agonists and antagonists. Nat Rev Drug Discov 1:847–858

10. Jamieson AG, Boutard N, Beauregard K et al (2009) Positional scanning for peptide secondary structure by systematic solid-phase synthesis of amino lactam peptides. J Am Chem Soc 131:7917–7927

11. Tatsu Y, Nishigaki T, Darszon A, Yumoto N (2002) A caged sperm-activating peptide that has a photocleavable protecting group on the backbone amide. FEBS Lett 525:20–24

12. Taniguchi A, Sohma Y, Kimura M et al (2006) "Click peptide" based on the "o-acyl isopeptide method": control of A beta1-42 production from a photo-triggered A beta1-42 analogue. J Am Chem Soc 128:696–697

13. Wood JS, Koszelak M, Liu J, Lawrence DS (1998) A caged protein kinase inhibitor. J Am Chem Soc 120:7145–7146

14. Kalbag SM, Roeske RW (1975) A photolabile protecting group for histidine. J Am Chem Soc 97:440–441

15. Pan P, Bayley H (1997) Caged cysteine and thiophosphoryl peptides. FEBS Lett 405:81–85

16. Ramesh D, Wieboldt R, Niu L et al (1993) Photolysis of a protecting group for the carboxyl function of neurotransmitters within 3 microseconds and with product quantum yield of 0.2. Proc Natl Acad Sci U S A 90:11074–11078

17. Barlos K, Chatzi O, Gatos D, Stavropoulos G (1991) 2-Chlorotrityl chloride resin. Studies on anchoring of Fmoc-amino acids and peptide cleavage. Int J Pept Protein Res 37:513–520

Chapter 4

Using Caged Ligands to Study Intracrine Endothelin Signaling in Intact Cardiac Myocytes

Clémence Merlen, Louis R. Villeneuve, and Bruce G. Allen

Abstract

Intracrine signaling refers to the activation of receptors located within the cell and many intracrine receptors have been localized to the nuclear membrane. The presence and function of nuclear receptors have been studied in isolated nuclei. Much less information is available concerning the function of these receptors within the context of intact cells due, in part, to difficulties in accessing the intracellular receptor without activating those at the cell surface. Here, we describe the use of caged agonists to study intracrine signaling in intact, living cells. The caging moiety permits cells to be loaded with a functionally "inert" ligand. After washing the cells free of extracellular caged ligand, a brief exposure to UV releases the native ligand within the cell. The actual duration of UV irradiation required is a function of the type of caging moiety employed and where it is incorporated into the ligand. Cells may then be assessed for changes in morphology, second messenger production, cellular signaling, or gene expression by confocal fluorescence microscopy, immunoblotting, or transcriptomic techniques.

Key words Caged ligands, GPCRs, Intracrine signaling, Immunocytofluorescence, Live-cell imaging, Confocal microscopy, Transcription

1 Introduction

Intracrine signaling refers to the activation of receptors located within the cell and many intracrine receptors have been localized to the nuclear membrane [1, 2]. The ligands for these receptors may be derived endogenously or taken up from the extracellular milieu. One of the difficulties in studying the molecular pharmacology or physiological function of intracellular receptors is selectively delivering agonists or antagonists to the receptor in intact cells and doing so with a degree of temporal and/or spatial control. Caged compounds are molecules such as nucleotides, ions, ligand-sensitive fluorescent dyes, or polypeptides that have been conjugated with an additional functional group. This group may be removed when desired by, for example, photolysis using UV

Bruce G. Allen and Terence E. Hébert (eds.), *Nuclear G-Protein Coupled Receptors*, Methods in Molecular Biology, vol. 1234, DOI 10.1007/978-1-4939-1755-6_4, © Springer Science+Business Media New York 2015

light or de-esterification by intracellular esterases. Generally, the caging moiety renders the ligand cell permeable. Depending upon the nature of the caging group and where it is incorporated on the ligand, the caged analog may also show reduced binding to the target receptor. Exposure to UV light removes the caging group, releasing the molecule in its original state. Since irradiation can be performed in a spatially and temporally controlled manner, caged compounds are valuable tools for the evaluation of intracellular events in living cells [3, 4]. Hence, caged compounds are widely used to study intracellular signaling, cellular metabolism, actin polymerization, gene regulation, and neurotransmission [5]. The first step is to introduce caged ligands into the cell. Several methods have been described to allow such compounds to enter cells: microinjection, bead loading, or choosing a cell-permeant caging group [4]. Bead loading and microinjection have certain limitations. Bead loading needs high concentrations of caged compounds and can only be employed on small numbers of cells. Microinjection permits the loading of only one cell at a time [5]. Moreover, both methods require the co-loading of inert fluorescent molecules such as fluorescein-dextran or rhodamine-dextran to evaluate the success of loading. Here, we described the use of cell-permeant caged ligands to activate or block the intracellular pool of receptors while sparing the cell surface receptors. Whereas bead loading and microinjection induce holes in plasma membrane, the use of cell-permeant compounds minimizes or avoids cell injury.

In this chapter, we describe two methods for uncaging ligands in function of the experiments to be performed: multi-cell versus single-cell. Single-cell experiments are well suited to assessments of changes in, for example, nuclear Ca^{2+} or NO content by live-cell imaging in real time. For single-live-cell experiments, cells are loaded with the caged ligand and uncaging is performed on the stage of a confocal microscope using a UV laser (i.e., 405 nm diode), allowing one to target specifically intracellular compartments as well as acquire data before, during, and after photolysis. Hence, rapid uncaging of the ligand occurs in the targeted area. Using this approach, we have examined the effects of intracellular release of endothelin-1 or isoproterenol upon nuclear calcium and NO levels using Fluo-4AM or DAF-2 DA, respectively [6, 7]. For experiments requiring intracellular ligand release on a larger scale, the use of a UV lamp mounted over a plate of cells allows release of active ligand simultaneously in a population of cells. In this latter case, one can then fix the cells for analysis by immunofluorescence, extract RNA, or lyse them for immunoblotting. Using these two approaches, we have determined that receptors associated with the nuclear membranes can regulate nuclear Ca^{2+} and nitric oxide signaling in cardiomyocytes [6, 7].

2 Materials

Prepare all solutions using ultrapure water (type 1; dH_2O) and analytical grade reagents wherever available. Prepare and store all reagents at room temperature (unless otherwise indicated).

1. Loading buffer: 10 mM HEPES, 134 mM NaCl, 6 mM KCl, 10 mM glucose, 2 mM $CaCl_2$, and 1 mM $MgCl_2$ (*see* **Note 1**). Mix, filter, and adjust to pH 7.4 with 1.0 M HCl. Use this solution at room temperature and store any unused loading buffer at 4 °C for up to 1 week.

2. Kruftbrühe (KB) buffer: 10 mM HEPES, 20 mM KCl, 10 mM KH_2PO_4, 70 mM K^+-glutamate, 100 mM K-aspartate, 2 mM $MgSO_4$, 5 mM creatine, 1 mM $MgCl_2$, 25 mM glucose, 10 mM β-hydroxybutyric acid, 20 mM taurine, 0.5 mM EGTA, and 0.1 % albumin. Mix, and adjust to pH 7.25 with KOH.

3. Fluo-4AM fluorescent cell-permeant calcium dye: For reconstitution, add 18 μL of 100 % DMSO to 50 mg of Fluo-4AM. Vortex and centrifuge for 1 min in a tabletop centrifuge. Then add 11.25 μL of pluronic acid F-127 (20 % solution in DMSO) and vortex for 3 min. Protect from light during reconstitution (*see* **Note 2**).

4. Phosphate-buffered saline (PBS): 137 mM NaCl, 2.7 mM KCl, 4.2 mM $Na_2HPO_4 \cdot H_2O$, 1.8 mM KH_2PO_4, pH 7.4 at room temperature.

5. DRAQ5 fluorescent cell-permeant DNA dye (5 mM solution): Protect from light (*see* **Note 2**).

6. Glass-bottomed 35 mm culture dishes.

7. High-intensity long-wave UV lamp (100 W, 365 nm, model B100AP lamp).

8. Cell culture plates (12 well, flat bottom).

9. Laminin from Engelbreth-Holm-Swarm murine sarcoma basement membrane (1 mg/mL): Dilute laminin to a concentration of 20 μg/mL with dH_2O. Coat the glass bottom of culture dishes with 300 μL of 20 μg/mL laminin solution. Incubate for 20 min at room temperature. Remove the remaining laminin solution and allow the plates to completely dry at room temperature before plating the cells.

10. Zeiss LSM 7 Duo microscope (combined LSM 710 and Zeiss Live systems) or comparable: The microscope stage (Zeiss Observer Z1) is equipped with a BC 405/561 dichroic mirror to allow simultaneous photolysis of the caged ligands and image acquisition (Zeiss Live). The system should be equipped with an objective heater system (37 °C) if an immersion objective is being used in conjunction with an environmental chamber

(37 °C, 5 % CO_2). The temperature above the objective and CO_2 levels within the environmental chamber should be monitored with an external device to verify that the desired conditions are being maintained. For example, the temperature of the objective heater may need to be increased in order to have the appropriate temperature at the objective level.

11. Caged ligand: In this case, we use caged endothelin-1 (cET-1) prepared as described previously [8].

3 Methods

3.1 Single-Cell Experiments

1. This protocol uses freshly isolated adult rat cardiac ventricular myocytes (ACVMs). The isolation of ACVMs has been described previously [9, 10].

2. Following isolation, suspend ACVMs in KB buffer. Plate 500 μL of this cell suspension (to attain a final confluence of around 50 %) on laminin-coated dishes and incubate for 1 h at 37 °C (*see* **Note 3**).

3. Gently aspirate the KB buffer with pipette and wash the cells by filling each dish with 2 mL of loading buffer at room temperature (*see* **Note 4**). Repeat for a total of three washes. After the last wash, remove any remaining loading buffer and add 1.5 mL of loading buffer.

4. Add 7.5 μL of the Fluo-4AM solution and then the caged ET-1 or vehicle. The concentration of caged ligand to be used in the loading buffer will have to be determined empirically (Fig. 1a; *see* **Note 5**). Gently agitate the dish in a back-and-forth motion to ensure mixing. Incubate for 25 min at room temperature. Avoid exposing the dish to direct light (*see* **Note 2**).

5. Add 1.5 μL of 5 mM DRAQ5 (for a final concentration of 1 μM) and incubate for at least 5 min at room temperature.

6. Gently aspirate the loading buffer and wash the cells with 2 mL of loading buffer for a total of three washes. Add 2 mL of loading buffer (*see* **Note 4**).

7. Incubate for 15 min at room temperature to allow deesterification of the Fluo-4AM.

8. Photolysis and live-cell imaging require a suitable instrument. We use a Zeiss LSM 7 Duo microscope with a 63×/1.4 oil Plan-Apochromat objective maintained at 37 °C using an objective heater system. Allow cells to equilibrate at 37 °C for 5 min before imaging.

9. Fluo-4 and DRAQ5 are excited and their fluorescence collected simultaneously using two different Zeiss Live detectors. To locate the nucleus, DRAQ5 is excited with a 635 nm/50 mW

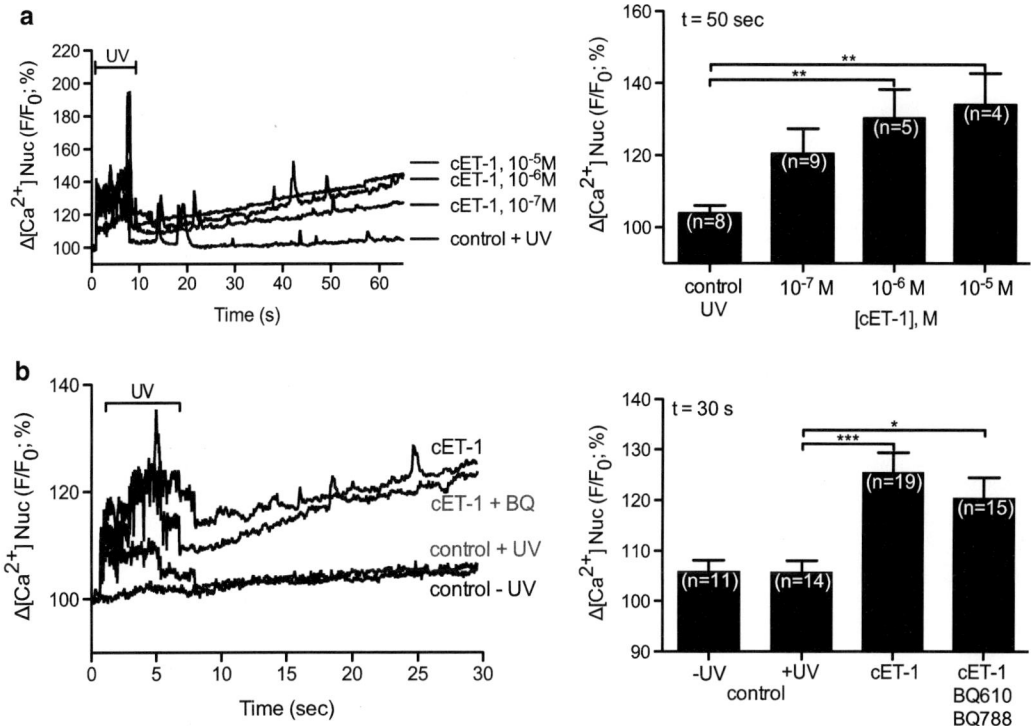

Fig. 1 Intracellular photolysis of a caged ET-1 analog induces an increase in nucleoplasmic $[Ca^{2+}]$. (**a**) Nucleoplasmic $[Ca^{2+}]$ recorded in rat ventricular myocytes before and after photolysis in cells preincubated with caged ET-1 at 10, 1, and 0.1 μM or vehicle. (**b**) Nucleoplasmic $[Ca^{2+}]$ recorded before and after photolysis in cells preincubated with 1 μM caged ET-1, caged ET-1 plus BQ610 (ETA antagonist), and BQ788 (ETB antagonist) (BQ; 1 μM each) or media alone (control). Controls were performed both with (control + UV) and without (control − UV) UV irradiation. DRAQ5 fluorescence was used to select the region corresponding to the nucleoplasm. Signals are presented as background-subtracted normalized fluorescence ($\%F/F_0$), where F is the fluorescence intensity, and F_0 is the resting fluorescence recorded in the same cell under steady-state conditions prior to photolysis. For each condition, the mean ± s.e.m. of nucleoplasmic Fluo-4 fluorescence at 30 or 50 s (as indicated) after photolysis is presented as a histogram. Number of cells is indicated in parentheses. *, $p < 0.05$; **, $p < 0.01$; ***, $p < 0.001$. Reprinted from ref. 6 with permission from Elsevier

diode and fluorescence emitted at >655 nm collected. Fluo-4 is excited with a 488 nm/100 mW diode (1–5 % laser intensity) and fluorescence emitted between 495 and 550 nm collected. Acquire about 200 frames (6.7 s) to establish a baseline for the Fluo-4 fluorescence emissions (F_0). To photolyze the cET-1, focus the 405 nm laser onto a 60 μm² rectangular region (*see* **Note 6**) overlapping the nucleus (*see* **Note 7**). Then photolyze the cET-1 using a 405 nm/30 mW diode (100 % laser intensity) for 7 s (*see* **Note 8**) while continuing to simultaneously acquire Fluo-4 and DRAQ5 fluorescence emissions (Fig. 2; *see* **Note 9**). Scan the cells at 30 fps in bidirectional mode (*see* **Note 10**). Set the pixel size at 0.2 μm and the pinhole at 2 Airy units (*see* **Note 11**).

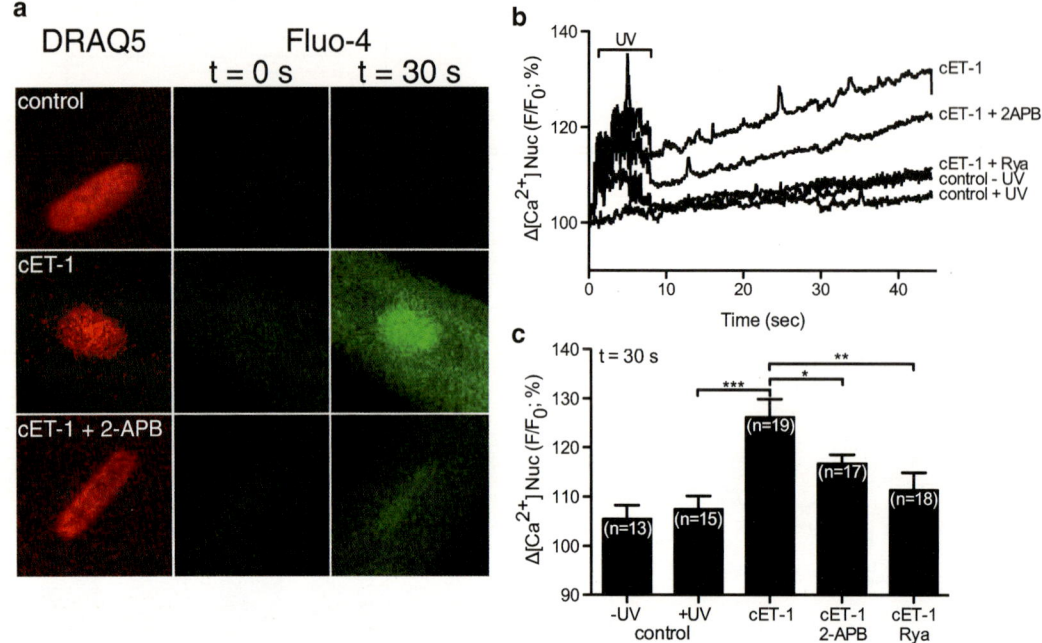

Fig. 2 Photolysis of caged ET-1 increases nucleoplasmic [Ca²⁺]. (**a**) Nucleoplasmic [Ca²⁺] recorded in rat ventricular myocytes before and after photolysis in cells preincubated with vehicle, 1 µM caged ET-1 (cET-1), or 1 µM caged ET-1 plus 20 µM of the IP₃R inhibitor, 2-APB. (**b**) Nucleoplasmic [Ca²⁺] recorded before and after photolysis in cells preincubated with 1 µM caged ET-1 (cET-1), caged ET-1 plus 20 µM 2-APB, ryanodine (Rya), or media alone (control). Controls were performed both with (control + UV) and without (control – UV) UV irradiation. DRAQ5 fluorescence was used to select the region corresponding to the nucleoplasm. Signals are presented as background-subtracted normalized fluorescence (%F/F₀), where F is the fluorescence intensity, and F₀ is the resting fluorescence recorded in the same cell under steady-state conditions prior to photolysis. (**c**) For each condition shown in *Panel b*, the mean ± s.e.m. of nucleoplasmic Fluo-4 fluorescence at 30 s after photolysis is presented as a histogram. Number of cells is indicated in parentheses. *, $p < 0.05$; **, $p < 0.01$; ***, $p < 0.001$. Reprinted from ref. 6 with permission from Elsevier

10. Several control conditions should be assessed (Figs. 1 and 2). These include "sham-loaded" cells to assess any effects of the photolysis procedure on the endpoint being measured. In addition, extracellular, non-permeable receptor antagonists should be employed to verify that the ligand has not diffused out of or been secreted by the cell and acted upon receptors at the cell surface (Fig. 1b). In addition, if available, using a caged (Fig. 3) [6] or cell-permeable antagonist [11] adds confidence that the phenomenon being observed is a result of activation of intracellular receptors.

11. Calculations: Following data acquisition, the changes in fluorescence within the nucleus, or other regions of interest (ROI) within the cell, over time can be quantified using the appropriate software (e.g., ImageJ, Volocity, Imaris, Zen). We delineate

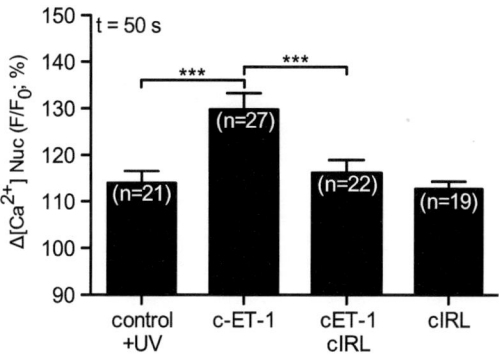

Fig. 3 A caged ETB antagonist (cIRL2500) inhibits the increase in $[Ca^{2+}]_n$ induced by caged ET-1. IRL2500 is a selective antagonist of the B-type endothelin receptor. Changes in nucleoplasmic $[Ca^{2+}]$ were recorded before and after photolysis in cells preincubated with or without caged ET-1 (cET-1) and in the presence or absence of caged IRL2500 (cIRL2500). DRAQ5 was used to target a region corresponding to the nucleoplasm. Signals are presented as background-subtracted normalized fluorescence ($\%F/F_0$), where F is the fluorescence intensity, and F_0 is the resting fluorescence recorded in the same cell under steady-state conditions prior to photolysis. For each condition, the mean ± s.e.m. of nucleoplasmic Fluo-4 fluorescence at 50 s is presented. Reprinted from ref. 6 with permission from Elsevier

the ROI and extract fluorescence intensities for each frame within Zen and then export the values to Excel. Fluorescence intensities (F) are then normalized to the fluorescence signal obtained prior to uncaging (F_0) using Excel. Normalized data is then exported to a graphics program and plotted as F/F_0 versus time. Data from multiple cells should be aligned according to the initiation of the 405 nm laser pulse (Figs. 1 and 2).

3.2 Multi-Celled Experiments

Experiments where assessment involves, for example, immunocytochemistry (Fig. 4), transcription, or immunoblotting, require the coordinated stimulation of cells in sufficient numbers to respect the detection limits of the respective endpoint. Immunocytochemistry experiments can be performed by uncaging entire populations of cells distributed as a monolayer in culture plates (*see* **Note 12**).

1. This protocol uses freshly isolated adult rat cardiac ventricular myocytes (ACVMs). The isolation of ACVMs has been described previously [9, 10].

2. Suspend freshly isolated cardiac myocytes in 2 mL KB buffer, plate 300 μL on laminin-coated cover slips, and incubate for 1 h at 37 °C in humid chamber. This should result in a final confluence of 80 %.

3. Using forceps transfer each cover slip into a separate well of a 12-well cell culture plate containing 1 mL of loading buffer.

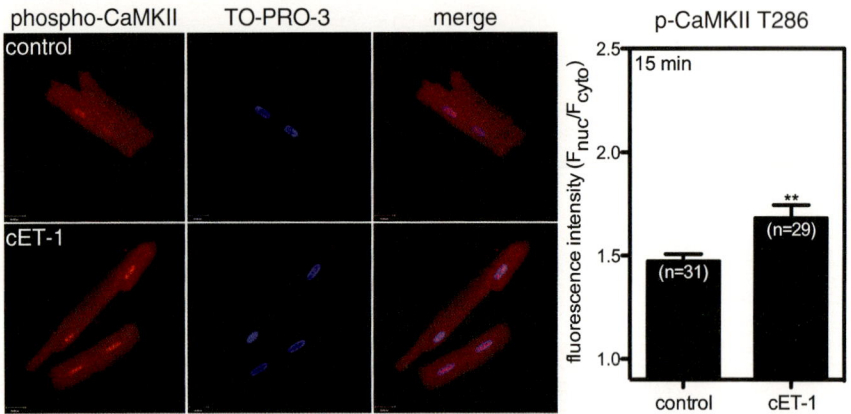

Fig. 4 Intracellular photolysis of a caged ET-1 analog induces activation of CaMKII within the nucleus. Autophosphorylation of CaMKII was studied by immunofluorescence in rat ventricular myocytes treated with cET-1 (1 μM) or vehicle. Ventricular myocytes were incubated with cET-1 for 30 min at room temperature. Following incubation, cells were washed, placed on ice, and exposed to a UV lamp for 3 min. Cells were then incubated at 37 °C for 15 min, fixed, and labeled with an antibody directed against phospho-CaMKII threonine-286 (*red*). Nuclei were stained with TO-PRO-3 (*blue*). Maximum intensity projection was performed to measure fluorescence intensity associated with phospho-CaMKII. Fluorescence intensity was determined in nuclei (Fnuc) and cytoplasm (Fcyto) and normalized to total area of the nucleus or cell. **, $p < 0.01$. Reprinted from ref. 6 with permission from Elsevier

Using a 1.5 mL plastic pipette, gently aspirate the buffer and refill each well with 1 mL of fresh loading buffer. Repeat for a total of three washes (*see* **Note 4**).

4. Gently aspirate the loading buffer and replace with 1 mL of loading buffer containing cET-1 or vehicle. Mix gently in a back-and-forth motion and incubate for 30 min at room temperature to allow loading of the cET-1 (*see* **Note 5**). Avoid exposing plate to direct light (*see* **Note 2**).

5. Following incubation, aspirate the loading buffer containing caged compounds. Wash each well three times with 1 mL of loading buffer (*see* **Note 4**).

6. Place the plates on ice to avoid heat from the UV lamp. Adjust the UV lamp such that the source is 3 cm above the plate. Expose the cells to UV light for 3 min. Make sure that the whole plate is uniformly exposed to the UV (*see* **Note 13**).

7. Following uncaging, incubate the cells at 37 °C for different times to allow biologically active ligand to exert their effect within the cells. At the end of the desired incubation time, stop the reaction on ice by fixing the cells for immunocytochemistry using 500 μL of PBS (pH 7.4) containing 2 % paraformaldehyde and 0.1 % Triton X-100 for 20 min (*see* **Note 12**).

4 Notes

1. The composition of the loading buffer must be adapted as a function of both the type of cell being studied and the fluorescent dyes employed in the experiment. The loading buffer described herein is that used with Fluo-4AM [6]. However, other indicator dyes may require modifications to the loading buffer. For example, when quantifying changes in NO production with DAF-2 DA, the loading buffer must be modified to 20 mM HEPES (from 10 mM) and supplemented with 1 % BSA [7]. Similarly, different cell types may require additional changes to the loading buffer. Avoid phenols, such as phenol red, in the loading buffer, as they may interfere with confocal imaging.

2. Protect dishes, plates, or tubes containing fluorescent dyes or caged compounds from direct light by covering the plate with aluminum foil to avoid bleaching or partial uncaging, respectively.

3. The incubations and washes described herein are those used with Fluo-4AM. However, these may need to be modified if other fluorescent dyes are to be used. For example, when quantifying changes in NO production with DAF-2 DA, there is an additional 30-min incubation at 4 °C prior to loading [7].

4. ACVMs are cylindrical in shape and thus do not present a large surface area for adhesion to the laminin. Care must be taken during mixing and washing so as not to dislodge and then inadvertently aspirate the cells. A gentle side-to-side movement is best during mixing. Addition and removal of buffers should be performed with care.

5. The incubation time and the concentration of the caged ligand in the loading buffer must be tested as a function of the type of cell being studied and the caged compound being used. Each compound will have unique cell permeability properties, resulting in different rates of uptake. The concentration of caged ligand required during loading will also depend on the dissociation constant (Kd) of ligand for its cognate receptor. Note that Kd of nuclear receptors may differ from that of the same receptor expressed at the cell surface. Hence, the optimal concentration of caged ligand employed during loading must be determined experimentally (Fig. 1). Finally, caged compounds may show limited solubility in aqueous solutions; thus stock solutions must be prepared by dissolving the compound in an organic solvent such as DMSO or methanol. Hence, controls must be performed to ensure that there is no effect of the vehicle used to initially dissolve these compounds (Figs. 1 and 2).

6. During uncaging, the 405 nm laser provides a point source of illumination that "scans" a designated area. Hence, uncaging

of the entire cell would affect the kinetics of the experiment as well as expose the cell to high levels of UV. To minimize UV damage, as well as provide a more rapid and spatially controlled release of ligand, the microscope is configured to scan a 60 μm^2 rectangular area adjacent to or overlapping the nucleus. The position of the nuclei (most ACVMs are binucleated) is revealed by the DRAQ5 fluorescence.

7. By placing the uncaging rectangle (*see* **Note 6**) at various distances from the nucleus, while quantifying Fluo-4 emissions within the nucleus, one may ascertain the ability of the uncaged ligand to diffuse to the nucleus.

8. The kinetics of photolysis may differ between caged compounds (e.g., *see* ref. 8).

9. During uncaging, the 405 nm laser pulse will temporarily increase the signal in Fluo-4 channel. This increase will last for the duration of the pulse.

10. The frequency (frame rate) and duration of image acquisition must be chosen as a function of the kinetics of the phenomena being observed. Avoid excessively high frame rates to minimize bleaching of the fluorescent dye (i.e., Fluo-4) or damaging the cells.

11. During the acquisition of fluorescence signals, it is important to periodically verify that the cells remain healthy and have not shifted position. Do not bump the instrument. Cells need to be healthy during the acquisition. If the morphology of cells changes during acquisition, it may be necessary to decrease the duration of UV exposure or the frame rate. One must find a balance between duration of exposure to UV that releases sufficient uncaged ligand while minimizing the possibility of cellular damage. A few seconds of exposure to UV should photolyze sufficient amounts of the caged compounds due to the intensity of the laser. Finally, insufficient de-esterification of the Fluo-4AM prior to initiating the experiment will result in an increase in the concentration of the de-esterified form of the dye during data acquisition and a continuous increase in fluorescence intensity.

12. The protocol as described permits analysis by immunocytofluorescence. For analysis of non-adherent cells by immunoblot or qPCR, the protocol may be modified. Following isolation, ACVMs can be incubated in tubes and then plated only during uncaging with the UV lamp (to ensure consistent and uniform exposure to the UV lamp). Following uncaging, the cells may then be transferred into tubes for washes and extraction.

13. The p38 MAP kinases are activated by cellular stresses, including exposure to UV [12]. No increase in p38 MAPK phosphorylation within the T-G-Y motif was detected in ACVMs following a 3-min exposure to the UV lamp.

Acknowledgements

This work was supported by grants from the Heart and Stroke Foundation of Quebec and the Canadian Institutes for Health Research (grant number MOP-64183, MOP-125970) to BGA. CM was the recipient of a Postdoctoral Fellowship from the Heart and Stroke Foundation of Canada.

References

1. Vaniotis G, Allen BG, Hébert TE (2011) Nuclear GPCRs in cardiomyocytes: an insider's view of β-adrenergic receptor signaling. Am J Physiol Heart Circ Physiol 301:H1754–H1764

2. Tadevosyan A, Vaniotis G, Allen BG et al (2012) G protein-coupled receptor signalling in the cardiac nuclear membrane: evidence and possible roles in physiological and pathophysiological function. J Physiol 590:1313–1330

3. McCray JA, Trentham DR (1989) Properties and uses of photoreactive caged compounds. Annu Rev Biophys Biophys Chem 18:239–270

4. Adams SR, Tsien RY (1993) Controlling cell chemistry with caged compounds. Annu Rev Plant Physiol Plant Mol Biol 55:755–784

5. Marriott G, Walker JW (1999) Caged peptides and proteins: new probes to study polypeptide function in complex biological systems. Trends Plant Sci 4:330–334

6. Merlen C, Farhat N, Luo X et al (2013) Intracrine endothelin signaling evokes IP3-dependent increases in nucleoplasmic Ca in adult cardiac myocytes. J Mol Cell Cardiol 62:189–202

7. Vaniotis G, Glazkova I, Merlen C et al (2013) Regulation of cardiac nitric oxide signalling by nuclear β-adrenergic and endothelin receptors. J Mol Cell Cardiol 62:58–68

8. Bourgault S, Létourneau M, Fournier A (2007) Development of photolabile caged analogs of endothelin-1. Peptides 28:1074–1082

9. Rodrigues B, Severson DL (1997) Preparation of cardiomyocytes. In: McNeill JH (ed) Biochemical techniques in the heart. CRC, New York, NY, pp 101–115

10. Chevalier D, Allen BG (2000) Two distinct forms of MAPKAP kinase-2 in adult cardiac ventricular myocytes. Biochemistry 39:6145–6156

11. Ryall KA, Saucerman JJ (2012) Automated imaging reveals a concentration dependent delay in reversibility of cardiac myocyte hypertrophy. J Mol Cell Cardiol 53:282–290

12. Dingar D, Merlen C, Grandy S et al (2010) Effect of pressure overload-induced hypertrophy on the expression and localization of p38 MAP kinase isoforms in the mouse heart. Cell Signal 22:1634–1644

Chapter 5

Quantification of Catecholamine Uptake in Adult Cardiac Myocytes

Erika F. Dahl, Casey D. Wright, and Timothy D. O'Connell

Abstract

In adult cardiac myocytes, multiple G protein-coupled receptors (GPCR) localize to and signal at the nucleus. These include endothelin B receptors, angiotensin type 1 and 2 receptors, β1- and β3-adrenergic receptors, and α1A- and α1B-adrenergic receptors. Initiation of signaling through nuclear GPCRs requires that ligands be produced within or transported into the cardiac myocytes, yet mechanisms whereby these ligands are produced or transported into cardiac myocytes are largely unclear. To activate nuclear adrenergic receptors in adult cardiac myocytes, uptake of endogenous catecholamines epinephrine and norepinephrine occurs via organic cation transporter 3 (OCT3), a member of the *slc22a* family of genes. This chapter details a method to detect and quantify catecholamine uptake in intact adult cardiac myocytes using a fluorescent-based catecholamine uptake assay.

Key words Cardiac myocyte, Adrenergic receptor, Catecholamine, Organic cation transporter

1 Introduction

1.1 Nuclear G Protein-Coupled Receptors (GPCRs)

As established in the preceding chapters, multiple GPCRs localize to the nuclei in adult cardiac myocytes, including endothelin B (ET$_B$R) receptors [1, 2], angiotensin type 1 (AT$_1$R) [3, 4] and type 2 receptors (AT$_2$R) [4], and adrenergic receptors (α1A-AR, α1B-AR, β1-AR and β3-AR) [5–9].

ET$_B$R: In sheep and adult rat ventricular myocytes, the ET$_B$R primarily localizes to and signals at the nuclei, whereas the endothelin A receptor (ET$_A$R) primarily localizes to and signals from the plasma membrane [1, 2]. Binding assays and immunoblotting using receptor-specific antibodies detect both the ET$_B$R and ET$_A$R at the nucleus; however in fractionated cardiac membranes only ET$_B$R immunoreactivity is detected on intracellular (nuclear) membranes [1, 2]. Functionally, a caged endothelin-1 (ET-1) that is activated intracellularly induces an inositol 1,4,5-trisphosphate (IP$_3$)-mediated increase in nuclear Ca^{2+} within seconds in adult rat

Bruce G. Allen and Terence E. Hébert (eds.), *Nuclear G-Protein Coupled Receptors*, Methods in Molecular Biology, vol. 1234, DOI 10.1007/978-1-4939-1755-6_5, © Springer Science+Business Media New York 2015

ventricular myocytes, which can be blocked by pretreating myocytes with a cell-permeable ET_BR-selective (IRL2500) antagonist [2].

AT₁R and AT₂R: In adult rat ventricular myocytes (ARVM), both the AT_1R and AT_2R localize to and signal at the nuclei [4]. Immunoblotting of fractionated cardiac myocytes with receptor-specific antibodies detects both the AT_1R and AT_2R in the nuclear fraction [4]. Angiotensin II induces *de novo* RNA synthesis within 30 min via both the AT_1R and AT_2R in adult rat ventricular myocytes, while the AT_1R induces Ca^{2+} transients within seconds in an IP_3-dependent manner during Ang-II stimulation in isolated nuclei [4].

β1- and β3-AR: In adult rat and mouse ventricular myocytes, both the β1- and β3-ARs localize to and signal at the nuclei [8, 9]. Radioligand binding assays detect β-ARs in enriched nuclear fractions isolated from adult cardiac myocytes, and immunocytochemistry with subtype-specific antibodies identifies both the β1- and β3-ARs at the nuclei [8]. However, the fraction of total β-ARs expressed at the nuclei is unclear [8]. Functionally, stimulation of β1- and β3-ARs in nuclei isolated from adult rat hearts produces distinctly different responses [8, 9]. In adult cardiac myocytes, β1-ARs activate adenylyl cyclase [8], likely through Gαs, whereas β3-AR stimulation activates nitric oxide production with seconds, likely through Gαi, and induces *de novo* transcription within 30 min [10].

α1-AR: In adult mouse ventricular myocytes (AMVM), which express the α1A and α1B, but not the α1D subtype, α1-ARs primarily localize to and signal at the nuclei. Radioligand binding assays on fractionated cardiac myocytes demonstrate that 80 % of α1-ARs segregate to the nuclear fraction [6]. Incubation of AMVM with BODIPY-prazosin, a fluorescent α1-AR ligand, identifies α1-ARs at the nucleus, but not at the plasma membrane [5–7]. Functionally, α1-AR signaling in adult cardiac myocytes requires both nuclear localization of the receptor, as mutation of nuclear localization sequences in α1-ARs disrupts signaling [7], and rapid (seconds) catecholamine uptake, as inhibition of catecholamine uptake blunts α1-signaling [6]. Further, CGP-12177A, a membrane-impermeable α1-AR antagonist, fails to block α1-AR signaling in AMVM, whereas prazosin, a membrane-permeable antagonist does, indicating a lack of functional α1-ARs at the plasma membrane [6].

1.2 Ligand Access to Nuclear GPCR

The identification of multiple GPCRs at the nuclei in adult cardiac myocytes raises the question of how signaling is activated, or, in simpler terms, how ligand might access a GPCR at the nuclear membrane. The "nuclear membrane" is a double-lipid bilayer consisting of the outer nuclear membrane (ONM) that is functionally contiguous with the endoplasmic reticulum (ER) (sarcoplasmic reticulum in myocytes), and the inner nuclear membrane (INM). The orientation of GPCRs in the nuclear membrane would impact how ligand accesses the receptor, how signaling is activated at the nucleus, and how signaling is transduced. Localization to the ONM would imply that signaling could originate at the nucleus, either in the lumen between the INM and ONM or in the cytoplasm, depending on orientation. Localization to the INM would imply that signaling could be initiated in the lumen between the INM and ONM or in the nucleoplasm. There is significant functional evidence indicating that in adult cardiac myocytes several nuclear GPCRs localize to the INM oriented with the C-terminus facing the nucleoplasm, thereby implying that nuclear GPCRs activate intranuclear signaling. As mentioned above, ETRs and β-ARs activate signaling in nuclei isolated from adult cardiac myocytes [1, 2, 8–11], and we have preliminary data suggesting that α1-ARs activate PKC in nuclei isolated from AMVM [12]. In total, these data suggest that nuclei from adult cardiac myocytes contain machinery sufficient to induce intranuclear GPCR signaling.

ETRs, ATRs, and ARs are activated by endogenous hormones (endothelin, angiotensin II, and norepinephrine/epinephrine) as well as by exogenous pharmaceuticals. To activate nuclear GPCRs these agents must be produced within or be transported into cardiac myocytes. Along those lines, a variety of mechanisms could be proposed to account for activation of nuclear GPCRs including (1) intracrine, synthesis of hormone within a cell leading to direct activation of the receptor; (2) autocrine, synthesis and secretion of hormone from a given cell and subsequent reuptake of the hormone; and (3) paracrine, synthesis and secretion of a hormone from one cell and uptake into a different cell. The latter two mechanisms, autocrine and paracrine, necessarily involve transport into the cell, either through passive diffusion or facilitated transport. Passive diffusion is defined as a concentration gradient-driven, mass transport of a molecule through the plasma membrane that is governed by the molecule's size and lipophilicity [13]. Facilitated transport involves the passage of molecules through particular integral membrane proteins down their electrochemical gradient.

The receptors for the peptide hormones endothelin and angiotensin II are likely activated by intracrine mechanisms, but activation through autocrine or paracrine mechanisms cannot be discounted.

ET_BR: Adult cardiac myocytes produce and secrete endothelin basally and in response to electrical stimuli, suggesting that intracrine-mediated activation of nuclear ET_BRs is possible. Further, cardiac fibroblasts from neonatal rat hearts secrete endothelin in response to angiotensin II, suggesting that paracrine activation of nuclear ET_BRs is possible, depending on ET-1 uptake [14].

AT-Rs: Neonatal rat ventricular myocytes exposed to high concentrations of glucose or isoproterenol produce angiotensin II that is detected in the nucleus [15]. Similarly, in diabetic adult rat hearts, high glucose increases intracellular angiotensin II [16].

Neonatal rat ventricular fibroblasts exposed to high concentrations of either glucose or isoproterenol also produce angiotensin II [17].

Therefore, for nuclear ET_BRs and ATRs, it is likely that intracellular pools of either endothelin or angiotensin II activate their respective receptors. In addition, trafficking of endothelin or angiotensin II might provide another mechanism to facilitate nuclear signaling. However, it is unclear whether a facilitated transport mechanism exists for either hormone, as is the case for endogenous catecholamines, epinephrine and norepinephrine, that are transported by members of the *slc22a* family of solute transporters as described below.

ARs: Organic cation transporters 1–3 (OCT1–3), encoded by the *slc22a* subfamily of solute transporters, transport organic cations, weak bases, and some neutral compounds acting as facilitative diffusion systems to transport substances in both directions across the plasma membrane [18]. All three OCTs mediate catecholamine transport [18–20], but show differential tissue-specific expression patterns [21–23]. OCT1 is expressed in the liver, kidney, and small intestine in rodents, whereas OCT1 is expressed primarily in the liver in humans [24]. OCT2 is expressed primarily in the kidney in both rodents and humans [25]. OCT3 is expressed primarily in the placenta, ovary, uterus, and heart in rodents, whereas OCT3 is expressed primarily in the liver, skeletal muscle, heart, and placenta in humans [26].

In AMVM, we found that OCT3 is expressed on both the plasma and nuclear membrane [6]. Using a fluorescent-based catecholamine uptake assay, the method we describe in detail in this chapter, we showed that catecholamines are transported into AMVM within seconds, are significantly increased by 5 min, and peak within 30 min [6]. We also found that blockade of OCT3 with corticosterone inhibits α1-AR-mediated phosphorylation of ERK1/2, suggesting a requirement for catecholamine uptake to activate nuclear α1-signaling [6]. Furthermore, hearts from OCT3KO mice show dramatically reduced cation uptake in vivo, but more importantly also show a trend toward reduced heart size in male but not female mice (WT: HW 160 ± 17 mg, OCT3KO: HW 145 ± 18 mg $n = 7$ $P = 0.138$) [27], similar to the small heart phenotype reported for male α1A- and α1B-AR double-knockout mice (WT: HW 147 ± 3 mg n=33, α1ABKO: HW 122 ± 3 mg $n = 27$ $P < 0.05$) [28].

The remainder of the chapter focuses on methods to detect catecholamine transport through OCT3 into adult cardiac myocytes, although the methods could easily be adapted to study any of the *slc22a* (OCT1–3) family members in a variety of cell types. Furthermore, we include methods to perform competition assays with various AR ligands, using norepinephrine as an example. Finally, by defining the function of OCT3, and possibly other *slc22a* family members, we might identify a novel drug target to modulate AR function in cardiac myocytes.

2 Materials

2.1 Recommended Microplate Readers

All recommended microplate readers are manufactured by Molecular Devices (Sunnyvale, CA). If the microplate reader being used is not listed here, refer to **Note 1** for microplate reader requirements. The protocol described here was designed for a 24-well plate.

1. FlexStation 3 Benchtop Multi-Mode Microplate Reader.
2. SpectraMax M5 Multi-Mode Microplate Reader.
3. Analyst GT.

2.2 Recommended Experimental Setup Parameters for Microplate Reader

The neurotransmitter transport uptake assay can be measured in either kinetic or endpoint mode allowing for high-throughput screening as well as mechanistic studies within the same cellular system. Table 1 shows the recommended experimental setup parameters for FlexStation, SpectraMax M5, Analyst GT, and Fusion Microplate Analyzer.

2.3 1× Perfusion Buffer

1× Perfusion buffer (1 L): 7.03 g NaCl, 1.1 g KCl, 0.082 g KH_2PO_4, 0.085 Na_2HPO_4, 0.3 g $MgSO_4$-$7H_2O$, 10 mL 1 M Na-HEPES, 0.39 g $NaHCO_3$, 3.75 g taurine, 1 g 2,3-butanedione, 1 g glucose.

For 1× perfusion buffer made fresh from powder at the time of use, dissolve ingredients in 990 mL of 18.2 MΩ H_2O and filter-sterilize. Adjust the pH to 7.0 with sterile HCl as needed. 1× perfusion buffer is dye free.

Table 1
Recommended experimental setup parameters for FlexStation, SpectraMax M5, and analyst GT

Parameters	FlexStation	SpectraMax M5	Analyst GT	Fusion microplate analyzer
Temperature	37 °C	37 °C	–	37 °C
Excitation wavelength (nm)	440	440	425–435	440
Emission wavelength (nm)	520	520	510–520	520
Emission cutoff (nm)	515	515	505 dichroic (bottom)	515
PMT sensitivity	Medium	Medium	–	Medium
Reads/well	6	6	–	6
Attenuator	–	–	Medium	–
Z-height	–	–	Bottom of plate	–

2.4 Stock Solutions

1. 1 mM Hydrochloric acid (HCl): 50 μL 1 M HCl, 50 mL 18.2 MΩ H_2O.

2. 100 mM L-ascorbic acid (AA): 88 mg AA, 5 mL 1 mM HCl. Store in 100 μL aliquots at –20 °C.

3. 10 mM Norepinephrine (NE): 16 mg NE, 5 mL 1 mM HCl. Store in 100 μL aliquots at –20 °C.

4. 200 μM NE: 20 μL 10 mM NE, 100 μL 100 mM AA, 880 μL 1× perfusion buffer.

2.5 Dye Solution

Whether using the Explorer or Bulk Neurotransmitter Transporter Uptake Assay Kit (Molecular Devices, Sunnyvale, CA), allow the vial containing lyophilized fluorescent dye/masking dye to equilibrate to room temperature.

1. Explorer kit: Reconstitute the lyophilized fluorescent dye/ masking dye by adding 10 mL of 1× perfusion buffer (*see* Subheading 2.3). Mix with a vortex until contents of vial are completely dissolved. This is the dye solution.

2. Bulk kit: Reconstitute the lyophilized fluorescent dye/masking dye by adding 10 mL of 1× perfusion buffer. Mix with a vortex until contents of vial are completely dissolved. Transfer the complete contents of the vial to a separate container and add another 90 mL 1× perfusion buffer to bring the total volume to 100 mL. This is the dye solution. For storage, *see* **Note 2**.

2.6 NE Working Stocks Prepared from 200 μM NE

1. 0.02 μM NE: 5 μL 200 μM NE, 50 mL 1× perfusion buffer.

2. 0.2 μM NE: 5 μL 200 μM NE, 5 mL 1× perfusion buffer.

3. 2 μM NE: 50 μL 200 μM NE, 5 mL 1× perfusion buffer.

4. 20 μM NE: 500 μL 200 μM NE, 5 mL 1× perfusion buffer.

2.7 Dilutions of NE Working Stocks

Final concentrations are based on the final concentration of NE in each well after adding 100 μL of each respective NE dilution to 500 μL of AMVM suspended in 1× perfusion buffer.

1. 0.3×10^{-3} μM NE: 7.5 μL 0.02 μM NE, 92.5 μL 1× perfusion buffer.

2. 1×10^{-3} μM NE: 25 μL 0.02 μM NE, 75 μL 1× perfusion buffer.

3. 3×10^{-3} μM NE: 7.5 μL 0.2 μM NE, 92.5 μL 1× perfusion buffer.

4. $1 \times 10\ 10^{-3}$ μM NE: 25 μL 0.2 μM NE, 75 μL 1× perfusion buffer.

5. 30×10^{-3} μM NE: 7.5 μL 2 μM NE, 92.5 μL 1× perfusion buffer.

6. 0.1 μM NE: 25 μL 2 μM NE, 75 μL 1× perfusion buffer.

7. 0.3 μM NE: 7.5 μL 20 μM NE, 92.5 μL 1× perfusion buffer.

8. 1 μM NE: 25 μL 20 μM NE, 75 μL 1× perfusion buffer.

9. 3 μM NE: 7.5 μL 200 μM NE, 92.5 μL 1× perfusion buffer.

10. 10 μM NE: 25 μL 200 μM NE, 75 μL 1× perfusion buffer.

11. 30 μM NE: 75 μL 200 μM NE, 25 μL 1× perfusion buffer.

3   Methods

3.1 Overview

Our purpose here is to measure the uptake of adrenergic receptor ligands into AMVM focusing on the uptake of the endogenous catecholamine norepinephrine (NE) using the following protocol, which is adapted from the Neurotransmitter Transporter Uptake Assay (Molecular Devices, Sunnyvale, CA). This technique allows for determination of the maximum catecholamine uptake rate (V_{max}) and transporter specificity using a fluorescent catecholamine analog and an unlabeled catecholamine competitor. Furthermore, this competition assay-based technique could also be adapted to determine rank order for the affinity and uptake of other compounds, both endogenous and exogenous, through catecholamine transporters. Defining the uptake kinetics and substrate affinity for OCT3 in AMVM may lead to novel therapies that regulate OCT3-mediated transport in the heart.

3.2 Preparation of Cardiac Myocytes and Reagents

1. *Isolation, Culture, and Seeding of Adult Mouse Ventricular Myocytes.* This protocol uses isolated adult ventricular myocytes, and we previously detailed methods for the isolation and culture of AMVM [29].

2. Prepare dye solution as detailed in Subheading 2.6.

3. If necessary, prepare compounds for competition assay (e.g., adrenergic agonists such as NE and isoproterenol, or adrenergic receptor antagonists such as prazosin and timolol). For NE, prepare a 200 μM NE working stock as detailed in Subheading 2.4.

4. For measurement of catecholamine uptake, we use 50,000 freshly isolated AMVM per assay condition with each assay condition measured in duplicate.

5. Suspend AMVM in 1× perfusion buffer and disperse in a 24-well tissue culture-treated clear, flat-bottom plate in a volume of 500 μL (*see* **Note 3**). Keep AMVM at 37 °C until needed for experimental assays.

6. To establish a competition curve, we use 11 different concentrations of an unlabeled competitor. For this protocol, NE is used as the unlabeled competitor at final concentrations of 0.3 nM, 1 nM, 3 nM, 10 nM, 30 nM, 100 nM, 300 nM, 1 μM, 3 μM, 10 μM, and 30 μM with one control well containing no

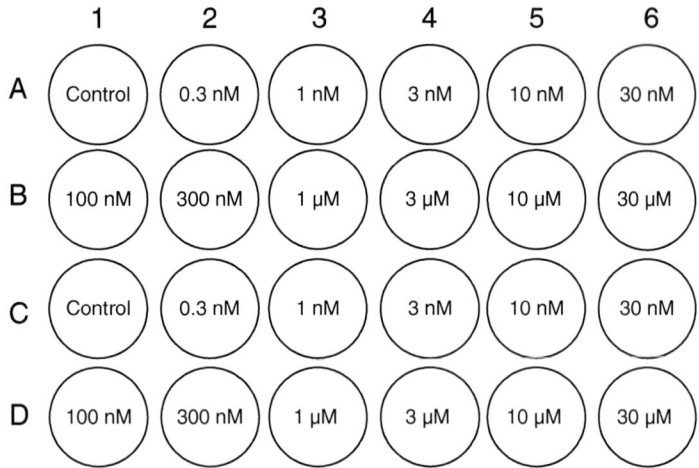

Fig. 1 Sample experimental competition assay setup using 11 different concentrations of competitor and 1 control well without competitor. The assay is performed in duplicate

competitor (*see* Subheadings 2.6 and 2.7). Therefore, using a 24-well plate will allow duplicate measurements for each dose, and will require a total of 1.2 million AMVM (24 wells × 50,000 AMVM/well = 1.2 million AMVM) as shown in Fig. 1.

3.3 Measurement of Catecholamine Uptake

1. At this point, cardiac myocytes have been seeded at 50,000 myocytes/well in 500 μL of 1× perfusion buffer in a 24-well tissue culture-treated clear, flat-bottom plate.

2. Add 100 μL of control or competitor (NE final concentration 0.3 nM–30 μM, *see* Fig. 1) and incubate for 15 min at 37 °C.

3. After incubation of the cells with the control or competitor, add 500 μL of dye solution (*see* Subheading 2.5) to each well and directly transfer plate to the microplate reader for kinetic analysis of catecholamine uptake (*see* **Note 4**).

4. Over a period of 15 min, measure each well in six distinct locations for 0.25 s each with a 30–70 % ratio of excitation to emission time.

3.4 Data Analysis

1. The kinetic parameters describing catecholamine uptake are determined by nonlinear curve fitting to generate a dose–response curve using a data analysis program such as Graph Pad PRISM 6.0 (Graph Pad, San Diego, CA) or any other suitable program.

2. Inhibition of catecholamine uptake by unlabeled NE is calculated from competition curves using a data analysis program, such as Graph Pad PRISM 6.0 or any other suitable program.

4 Notes

1. The microplate reader must have a bottom-read fluorescence optical system, ability to read 24-well plates (for this assay, although other cell types could work in smaller well sizes), and a kinetics mode.

2. Reconstituted dye solution is stable for up to 8 h at room temperature. Aliquots can be frozen at −20 °C and stored for up to 5 days without loss of activity.

3. In our experience, myocytes should not be placed in smaller wells (96-well plate for example). To ensure even distribution and avoid aggregation of AMVM in the wells, slide plates back and forth and side to side rather than swirling.

4. This step needs to be done quickly and use of a multichannel pipette is recommended.

References

1. Boivin B, Chevalier D, Villeneuve LR et al (2003) Functional endothelin receptors are present on nuclei in cardiac ventricular myocytes. J Biol Chem 278:29153–29163

2. Merlen C, Farhat N, Luo X et al (2013) Intracrine endothelin signaling evokes IP3-dependent increases in nucleoplasmic Ca in adult cardiac myocytes. J Mol Cell Cardiol 37:189–202

3. Lee DK, Lanca AJ, Cheng R et al (2004) Agonist-independent nuclear localization of the apelin, angiotensin AT1, and bradykinin B2 receptors. J Biol Chem 279:7901–7908

4. Tadevosyan A, Maguy A, Villeneuve LR et al (2010) Nuclear-delimited angiotensin receptor-mediated signaling regulates cardiomyocyte gene expression. J Biol Chem 285:22338–22349

5. Huang Y, Wright CD, Merkwan CL et al (2007) An alpha1A-adrenergic-extracellular signal-regulated kinase survival signaling pathway in cardiac myocytes. Circulation 115:763–772

6. Wright CD, Chen Q, Baye NL et al (2008) Nuclear alpha1-adrenergic receptors signal activated ERK localization to caveolae in adult cardiac myocytes. Circ Res 103:992–1000

7. Wright CD, Wu SC, Dahl EF et al (2012) Nuclear localization drives alpha1-adrenergic receptor oligomerization and signaling in cardiac myocytes. Cell Signal 24:794–802

8. Boivin B, Lavoie C, Vaniotis G et al (2006) Functional beta-adrenergic receptor signalling on nuclear membranes in adult rat and mouse ventricular cardiomyocytes. Cardiovasc Res 71:69–78

9. Vaniotis G, Del Duca D, Trieu P et al (2011) Nuclear beta-adrenergic receptors modulate gene expression in adult rat heart. Cell Signal 23:89–98

10. Vaniotis G, Glazkova I, Merlen C et al (2013) Regulation of cardiac nitric oxide signaling by nuclear beta-adrenergic and endothelin receptors. J Mol Cell Cardiol 37:58–68

11. Bkaily G, Choufani S, Sader S et al (2003) Activation of sarcolemma and nuclear membranes ET-1 receptors regulates transcellular calcium levels in heart and vascular smooth muscle cells. Can J Physiol Pharmacol 81:654–662

12. Wu SC, Wright CD, Cypher AL et al (2012) Nuclear targeting of the alpha 1A-adrenergic receptor is required for cardiac myocyte contractility. Circ Res 111:A128

13. Sugano K, Kansy M, Artursson P et al (2010) Coexistence of passive and carrier-mediated processes in drug transport. Nat Rev Drug Discov 9:597–614

14. Gray MO, Long CS, Kalinyak JE et al (1998) Angiotensin II stimulates cardiac myocyte hypertrophy via paracrine release of TGF-β1 and endothelin-1 from fibroblasts. Cardiovasc Res 40:352–363

15. Singh VP, Le B, Bhat VB et al (2007) High-glucose-induced regulation of intracellular ANG II synthesis and nuclear redistribution in cardiac myocytes. Am J Physiol Heart Circ Physiol 293:H939–H948

16. Singh VP, Le B, Khode R et al (2008) Intracellular angiotensin II production in diabetic rats is correlated with cardiomyocyte

apoptosis, oxidative stress, and cardiac fibrosis. Diabetes 57:3297–3306

17. Singh VP, Baker KM, Kumar R (2008) Activation of the intracellular renin-angiotensin system in cardiac fibroblasts by high glucose: role in extracellular matrix production. Am J Physiol Heart Circ Physiol 294:H1675–H1684

18. Nies AT, Koepsell H, Damme K et al (2011) Organic cation transporters (OCTs, MATEs), in vitro and in vivo evidence for the importance in drug therapy. Handb Exp Pharmacol 201: 105–167

19. Hayer-Zillgen M, Bruss M, Bonisch H (2002) Expression and pharmacological profile of the human organic cation transporters hOCT1, hOCT2 and hOCT3. Br J Pharmacol 136: 829–836

20. Roth M, Obaidat A, Hagenbuch B (2012) OATPs, OATs and OCTs: the organic anion and cation transporters of the SLCO and SLC22A gene superfamilies. Br J Pharmacol 165:1260–1287

21. Grundemann D, Gorboulev V, Gambaryan S et al (1994) Drug excretion mediated by a new prototype of polyspecific transporter. Nature 372:549–552

22. Jonker JW, Wagenaar E, Mol CA et al (2001) Reduced hepatic uptake and intestinal excretion of organic cations in mice with a targeted disruption of the organic cation transporter 1 (Oct1 [Slc22a1]) gene. Mol Cell Biol 21: 5471–5477

23. Schweifer N, Barlow DP (1996) The Lx1 gene maps to mouse chromosome 17 and codes for a protein that is homologous to glucose and polyspecific transmembrane transporters. Mamm Genome 7:735–740

24. Gorboulev V, Ulzheimer JC, Akhoundova A et al (1997) Cloning and characterization of two human polyspecific organic cation transporters. DNA Cell Biol 16:871–881

25. Grundemann D, Koster S, Kiefer N et al (1998) Transport of monoamine transmitters by the organic cation transporter type 2, OCT2. J Biol Chem 273:30915–30920

26. Koepsell H, Lips K, Volk C (2007) Polyspecific organic cation transporters: structure, function, physiological roles, and biopharmaceutical implications. Pharm Res 24:1227–1251

27. Zwart R, Verhaagh S, Buitelaar M et al (2001) Impaired activity of the extraneuronal monoamine transporter system known as uptake-2 in Orct3/Slc22a3-deficient mice. Mol Cell Biol 21:4188–4196

28. O'Connell TD, Ishizaka S, Nakamura A et al (2003) The alpha(1A/C)- and alpha(1B)-adrenergic receptors are required for physiological cardiac hypertrophy in the double-knockout mouse. J Clin Invest 111:1783–1791

29. O'Connell TD, Rodrigo MC, Simpson PC (2007) Isolation and culture of adult mouse cardiac myocytes. Methods Mol Biol 357: 271–296

Chapter 6

Characterization of the Interaction Between the Prostaglandin D₂ DP1 Receptor and the Intracellular L-Prostaglandin D Synthase

Chantal Binda and Jean-Luc Parent

Abstract

Identification of G protein-coupled receptor (GPCR)-interacting proteins is an intense subject of current research. However, confirmation and characterization of identified interactions can be difficult with GPCRs, especially at the endogenous level. Here, we describe how we characterized the interaction between the prostaglandin D₂ DP1 receptor and the intracellular L-type prostaglandin D synthase by in vitro pull-down assays using purified recombinant GST- and His-tagged proteins, by co-immunoprecipitation of overexpressed Flag- and HA-tagged proteins, and by co-immunoprecipitation of endogenous proteins.

Key words GPCR, GST pull-down, Co-immunoprecipitation, Endogenous

1 Introduction

GPCRs are, with over 800 members, among the most abundant membrane proteins in humans [1]. They respond to a plethora of extracellular ligands, initiating their intracellular responses. The importance of GPCRs in the progression and treatment of various diseases is highlighted by the fact that at least 30–40 % of prescription drugs target these proteins [2, 3]. Their potential for additional therapeutic benefits is considerable: currently available drugs target pathways that are controlled by only 10 % of the known GPCRs [2, 4]. GPCRs are accompanied along their life cycle by a range of interacting proteins to assist nascent receptors in proper folding, to target them to appropriate subcellular compartments, and to fulfill their signaling tasks, or alternatively to target them for degradation. Interfering with GPCR interaction networks is a promising strategy for specific therapeutic intervention [5–7]. It is thus not surprising to see such intense research efforts put into identification of GPCR-interacting proteins. Various approaches have been used to identify GPCR-interacting proteins such as yeast two-hybrid

Bruce G. Allen and Terence E. Hébert (eds.), *Nuclear G-Protein Coupled Receptors*, Methods in Molecular Biology, vol. 1234, DOI 10.1007/978-1-4939-1755-6_6, © Springer Science+Business Media New York 2015

screening using intracellular domains of the receptors as bait, as well as epitope-tag and TAP-tag purification of the receptors followed by mass spectrometry analyses of the proteins found in receptor complexes. Confirmation and characterization of the identified interactions with intracellular proteins can however be a challenge with GPCRs, given their multiple membrane-spanning and intracellular domains, their tendency to aggregate during solubilization, and, except for a few members of the family, their poor ability to be immunoprecipitated. This is especially true when one is trying to confirm the interactions at the endogenous context due to the low expression levels of GPCRs and often to the lack of good antibodies to immunoprecipitate endogenous receptors. Here, using the recently reported interaction between the DP1 PGD_2 receptor (DP1) and the intracellular synthase for its agonist, L-prostaglandin synthase (L-PGDS) [8], we describe the techniques used to confirm and characterize the interaction between a GPCR and an intracellular interacting protein by in vitro pull-down assays to demonstrate direct interaction and by co-immunoprecipitation between the GPCR and the interacting protein in conditions of overexpression and endogenous expression.

2   Materials

2.1   Protein Purification

1. LB (Luria Broth): 10 % (w/v) tryptone, 5 % (w/v) yeast extract, 10 % (w/v) NaCl (EMD Chemicals, Cat. # 71753-5).

2. Overexpress™ C41 (DE3) *Escherichia coli* strain (Avidis).

3. IPTG [Isopropyl β-D-thiogalactopyranoside] (Promega, Cat. # V395A).

4. Lysozyme (Sigma, Cat. # L6876-1G).

5. Glutathione Sepharose beads (GE Healthcare, Cat. # 17-0780-01).

6. Ni-NTA agarose beads (Qiagen, Cat. # 1018244).

7. Bovine Serum Albumin (Wisent, Cat. # 800-095-CG).

8. Protease inhibitors (CLAP): 10 nM chymostatin, 10 nM leupeptin, 9 nM antipain, and 9 nM pepstatin.

9. Binding buffer: 150 mM NaCl, 10 mM Tris–HCl (pH 7.4), 1 mM EDTA, 10 % glycerol, 0.5 % IGEPAL, 1:500 CLAP.

10. Lysis buffer: 300 mM NaCl, 50 mM NaH_2PO_4, 10 mM imidazole, 1 % sarkosyl, 1:500 CLAP, pH 8.0.

11. Washing buffer: 300 mM NaCl, 50 mM NaH_2PO_4, 20 mM imidazole, 1:500 CLAP, pH 8.0.

12. Elution buffer: 300 mM NaCl, 50 mM NaH_2PO_4, 250 mM imidazole, 1:500 CLAP, pH 8.0.

13. PBS (phosphate-buffered saline) buffer: 8 g NaCl, 0.2 g KCl, 1.44 g Na_2HPO_4, 0.2 g KH_2PO_4. Adjust to 1 L with dH_2O.

14. 4× SDS-PAGE sample buffer: 250 mM Tris–HCl pH 6.8, 40 % (v/v) glycerol, 8 % (w/v) SDS, 4 % (v/v) β-mercaptoethanol, 10 ng bromophenol blue or to desired darkness.

15. Coomassie Blue Staining solution: 0.4 % (w/v) Coomassie Brilliant Blue, 50 % (v/v) ethanol, 10 % (v/v) acetic acid.

16. Destaining solution: 15 % (v/v) ethanol, 5 % (v/v) acetic acid.

17. SDS-PAGE gel (10 % acrylamide).

18. Western blot transfer buffer: 12 g Tris base, 57.6 g glycine, 20 % methanol in final volume of 4 L.

19. TBS-T buffer: 20 mM Tris–HCl pH 8.0, 150 mM NaCl, 0.1 % Tween-20.

20. Blocking solution: 4 % nonfat dry milk in TBS-T.

21. Primary antibodies for immunoblotting: α-GST HRP-conjugated antibody (goat, Bethyl Laboratories) or α-L-PGDS polyclonal antibody (rabbit, Cayman Chemical Co.) (diluted 1:30,000 and 1:1,000, respectively, in 4 % nonfat milk made in TBS-T).

22. Secondary antibody: Anti-rabbit horseradish peroxidase-linked HRP (GE Healthcare) (diluted 1:10,000 in 4 % nonfat milk made in TBS-T).

23. ECL (GE Healthcare).

2.2 Cell Culture

1. 60 mm dishes (Adherent).

2. Cell culture media: Dulbecco's modified Eagle's medium (DMEM) (Wisent), McCoy's 5A medium (Wisent).

3. Fetal bovine serum (FBS) (Gibco).

4. Stimulation medium: DMEM, 0.5 % BSA, 20 mM Hepes.

5. Human embryonic kidney 293 (HEK293) cells.

6. Human colorectal adenocarcinoma (HT29) cells.

2.3 Cell Transfection

1. Transfection reagent: *Trans*IT®-LT1 reagent (Mirus).

2. Opti-MEM medium (Gibco).

3. pcDNA3 plasmid (Invitrogen), pcDNA3-L-PGDS-HA, pcDNA3-FLAG-DP1. The FLAG-DP1 and L-PGDS-HA DNA were amplified by PCR, digested with BamHI and EcoRI, and ligated into the pcDNA3 vector digested with the same enzymes as described in Binda et al. [8].

2.4 Cell Lysis

1. Sterile PBS.

2. Prostaglandin D_2 (PGD_2) (Cayman Chemical Co.).

3. PBS (stored at 4 °C).

4. Lysis buffer: (150 mM NaCl, 50 mM Tris (pH 8.0), 0.5 % deoxycholate, 0.1 % SDS, 10 mM $Na_4P_2O_7$, 1 % IGEPAL, 5 mM EDTA) supplemented with protease inhibitors (CLAP). Store at 4 °C.

5. Bio-Rad protein assay: Protein assay reagent A (Cat. # 500-0113), Protein assay solution B (Cat. # 500-0114), Protein assay solution C (Cat. # 500-0115).

6. 96-Well plate (Sarstedt).

7. 96-Well plate spectrophotometer reader (for 595 nm).

2.5 Immunopre-cipitation

1. Protein G-agarose beads (Santa Cruz Biotechnology, Cat. # sc-2001).

2. SDS-PAGE gel (10 % acrylamide).

3. Western blot transfer buffer (12 g Tris base, 57.6 g glycine, 20 % methanol in final volume of 4 L).

4. TBS-T buffer: 20 mM Tris–HCl pH 8.0, 150 mM NaCl, 0.1 % Tween-20.

5. Blocking solution (4 % nonfat dry milk in TBS-T).

6. Primary antibody for immunoprecipitation: α-FLAG M_2 (mouse, Sigma-Aldrich) and α-L-PGDS monoclonal antibody (mouse, Cayman Chemical Co.).

7. Primary antibodies for immunoblotting: α-HA-peroxidase high affinity antibody (HA-HRP) (Roche Applied Science, Cat. # 3F10), α-FLAG polyclonal antibody (rabbit, Sigma-Aldrich), α-L-PGDS polyclonal antibody (rabbit, Cayman Chemical Co.), α-DP1 polyclonal antibody (rabbit, Cayman Chemical Co.) (diluted 1:2,000, 1:1,000, 1:1,000, 1:1,000, respectively, in 4 % nonfat milk made in TBS-T).

8. Secondary antibody: Anti-rabbit horseradish peroxidase-linked HRP (GE Healthcare) (diluted 1:10,000 in 4 % nonfat milk made in TBS-T).

9. ECL reagent (GE Healthcare).

3   Methods

3.1 In Vitro Pull-Down Binding Assays

Protein purification protocol was adapted from Cold Spring Harbor Protocols [9].

3.1.1 Production of Recombinant GST/His-Tagged Proteins

The ICL1 (amino acids 46–57) and ICL2 (amino acids 129–150) fragments of DP1 were generated by annealing two pairs of complementary oligonucleotides. Equal quantities of oligonucleotides were mixed and denatured by boiling for 5 min. The mix was then incubated at room temperature for 30 min to allow hybridization, digested with EcoRI and XhoI, and ligated into the pGEX4T-1 vector digested with the same enzymes. The ICL3 (amino acids 217–262) and C-tail (CT) (amino acids 332–359) fragment of DP1 were amplified by PCR, digested with EcoRI and XhoI, and ligated into the pGEX4T-1 vector digested with the same enzymes. To produce His-tagged proteins, PCR fragments corresponding to the cDNA coding for full-length L-PGDS was amplified by PCR, digested with BamHI and EcoRI, and inserted into the pRSETA expression vector digested with the same enzymes.

3.1.2 Purification of Recombinant GST/ His-Tagged Proteins

1. Production of recombinant proteins using Overexpress™ C41 (DE3) *Escherichia coli* strain (Avidis) (*see* **Note 1**, Fig. 1). Thaw an aliquot of competent cells on ice. Add 1–2 µg of pRSETA-L-PGDS, pGEX4T-1-DP1-ICL1, pGEX4T-1-DP1-ICL2, pGEX4T-1-DP1-ICL3, and pGEX4T-1-DP1-CT (*see* **Note 2**) to a 50 µL aliquot of C41 cells for each construct. Mix gently by inverting the tube and incubate on ice for about 10–20 min. Place the tube at 37 °C for 5 min, and then place it back on ice for 2 min. Add 200 µL of LB medium and incubate the tube for 1 h at 37 °C. After 1 h, pour the tube content in 100 mL of LB medium containing ampicillin (diluted 1:1,000) (*see* **Note 3**) and incubate at 37 °C overnight with constant agitation.

2. Transfer 1 mL of the overnight bacterial culture into 100 mL of LB medium containing the appropriate antibiotic and incubate at 37 °C. Grow until the culture reaches an OD_{595} value between 0.6 and 1.0 (*see* **Note 4**). Put the culture on ice for 10 min and induce with 0.4 mM IPTG and 2 % ethanol at room temperature overnight with constant agitation (*see* **Notes 5** and **6**).

3. Centrifuge the culture at $3,200 \times g$ for 20 min at 4 °C. Remove supernatant and resuspend the pellet in 1 mL of binding buffer per 50 mL of culture for GST-tagged proteins or 1 mL of lysis buffer with 1 % sarkosyl per 50 mL of culture for $(His)_{6\text{-tagged}}$ proteins (*see* **Note 7**). Incubate for 30 min at 4 °C with rotation in the presence of 1 mg/mL of lysozyme.

4. In parallel, take 100 µL of glutathione Sepharose beads or 150 µL of Ni-NTA beads per 100 mL of initial culture, mark the total volume (*see* **Note 8**), and wash three times with 1 mL of $1\times$ PBS containing 10 mg/mL BSA (spin: 1 min at $375 \times g$). Adjust the bead volume with the same buffer to the initial mark and incubate for 30 min at room temperature with rotation.

5. Sonicate the samples for 20 s at an intensity of 2–3 (*see* **Note 9**). Repeat sonication four times, alternating for 20 s on ice between each sonication.

6. Centrifuge the samples at $16,000 \times g$ for 20 min at 4 °C and keep the supernatants (*see* **Note 10**). For the $(His)_6$-tagged

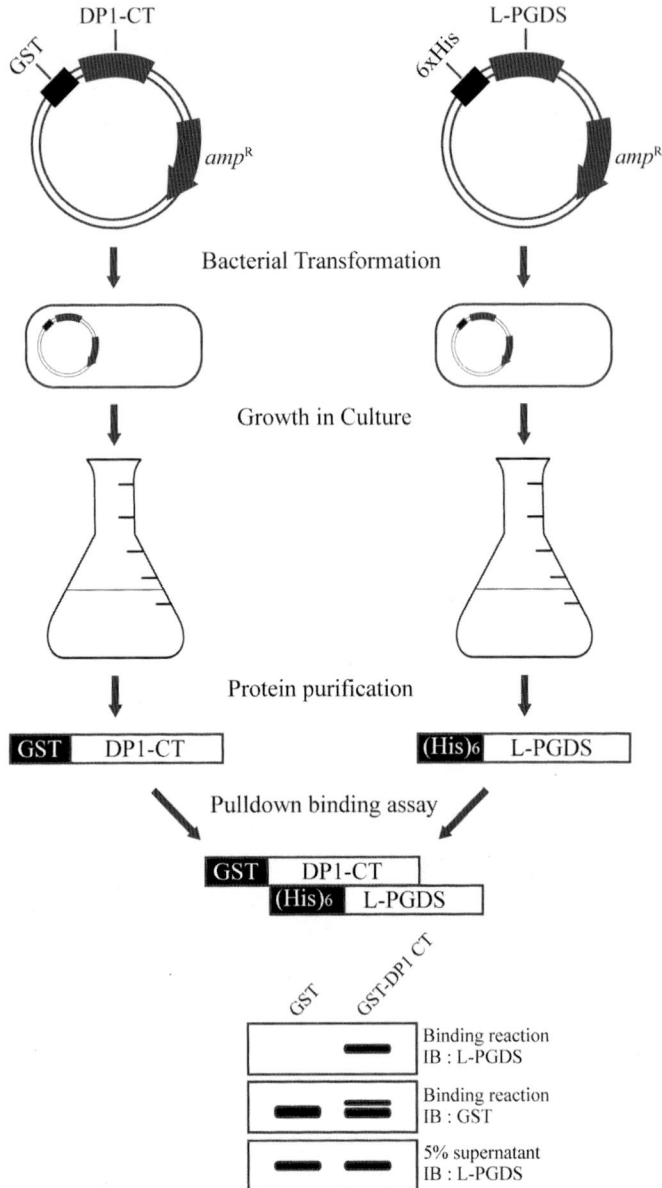

Fig. 1 Protein purification and pull-down binding assay. Briefly, the cDNA of each protein of interest is inserted into the appropriate expression vectors. Protein purification is then carried out following the protocol mentioned in this chapter before proceeding to the pull-down binding assays. These assays are then analyzed by SDS-PAGE gel and Western blotting. In addition to the GST fusion protein of interest, a GST band is often seen when blotting with the anti-GST antibody, as illustrated in the GST-DP1-CT lane

proteins, add 2 % Triton X-100 to the supernatants and incubate for 30 min at 4 °C with rotation (*see* **Note 11**).

7. Combine supernatants with the previously washed beads from step four and incubate for 1 h at 4 °C with rotation.

Use glutathione Sepharose beads for GST-tagged proteins and Ni-NTA beads for the $(His)_6$-tagged proteins.

8. Wash five times the beads with 1 mL binding buffer for GST-tagged proteins and with washing buffer for the $(His)_6$-tagged proteins (spin: 1 min at $375 \times g$).

9. Resuspend the glutathione Sepharose beads with 0.2 mL of binding buffer containing 20 % glycerol (*see* **Note 12**). Elute the $(His)_6$-tagged proteins four times by using the elution buffer (spin: 1 min at $375 \times g$). Use 500 µL for the first elution and 250 µL for the subsequent three: let the beads incubate in elution buffer for 5 min between each elution. Add 30 % glycerol to each elution sample collected. Store all the purified proteins at −80 °C (*see* **Notes 13** and **14**).

10. Analyze purified recombinant proteins by SDS-PAGE. Stain the gel with Coomassie Brilliant Blue R-250 for 30 min with agitation. Destain the gel with destaining solution overnight with agitation. Purified proteins should be colored in blue.

3.1.3 Pull-Down Binding Assay

1. Take 5 µg of the purified recombinant GST-tagged proteins and wash twice with 1 mL of binding buffer containing 1 mM dithiothreitol (DTT) (*see* **Note 15**, Fig. 1).

2. Add 5 µg of the purified recombinant $(His)_6$-tagged proteins to the GST-tagged proteins and complete the volume to 200 µL with binding buffer plus 1 mM DTT. Incubate overnight at 4 °C with rotation (*see* **Note 16**).

3. Spin binding reactions for 1 min at $375 \times g$. Keep an aliquot of 20 µL of the supernatant for SDS-PAGE analysis and add 10 µL of 4× SDS-PAGE sample buffer. Aspirate the rest of the supernatant and wash three times with binding buffer plus 1 mM DTT. Add 40 µL of 1× SDS-PAGE sample buffer to the beads. Heat samples at 100 °C for 5 min, spin for 1 min at $375 \times g$, and proceed with SDS-PAGE analysis.

4. Prepare three SDS-PAGE gels. On the first gel, load 5 µL of the binding reactions. Load 25 µL of the binding reactions on gel two (*see* **Note 17**). Finally, load 25 µL of supernatants on gel three. After electrophoresis, transfer to nitrocellulose membranes (100 V, 60 min, Minigel transfer cell).

5. Block membranes with blocking solution for 15 min at room temperature with gentle agitation.

6. Incubate the membrane from first gel with α-GST-HRP antibody (diluted 1:30,000 in blocking solution) and the other membranes with polyclonal α-L-PGDS antibody (diluted 1:1,000 in 4 % blocking solution) for 1 h at room temperature with gentle agitation. Alternatively, an α-histidine antibody can be used for detection of the His-tagged interacting partner (Cell Signaling Technology, diluted 1:2,000 in blocking solution).

7. Wash membranes three times for 8 min each with TBS-T.

8. Add α-rabbit secondary antibody to membranes 2 and 3 (diluted 1:10 000 in blocking solution) and incubate for 1 h at room temperature with gentle agitation (*see* **Note 18**).

9. Wash membranes three times for 8 min each with TBS-T.

10. Add ECL solution, incubate, and expose to film.

3.2 G Protein-Coupled Receptor Immunoprecipitation (Heterologous Expression)

3.2.1 Cell Culture

1. Seed 7×10^5 HEK293 cells in 60 mm dishes in 5 mL DMEM supplemented with 10 % FBS per condition.

2. Incubate at 37 °C, 5 % CO_2, for 24 h.

3.2.2 Cell Transfection

1. Prepare one microtube for DNA per condition. Here, conditions used were pcDNA3, pcDNA3-L-PGDS-HA, pcDNA3-FLAG-DP1, and pcDNA3-L-PGDS-HA + pcDNA3-FLAG-DP1 (*see* **Note 19**).

2. Prepare a stock of transfection reagent (*Trans*IT®-LT1) using a ratio of 2 μL of *Trans*IT®-LT1 for 1 μg of DNA per condition in opti-MEM (200 μL per condition) for all your conditions in a Falcon tube.

3. Add the appropriate amount of DNA in each DNA tube per condition. Here, 5 μg of L-PGDS-HA and 0.05 μg of FLAG-DP1 were added to DNA tubes depending on conditions for a total amount of 5.05 μg of DNA transfected (*see* **Notes 20–22**).

4. Add 200 μL of the transfection reagent mix to each DNA microtube. Mix gently the DNA and transfection reagent by carefully pipetting up and down a couple of times. Incubate at room temperature for 20 min.

5. Add the mix from **step 4** drop by drop in the 60 mm dishes. Homogenize by gently rocking the dishes and incubate at 37 °C, 5 % CO_2, for 48 h.

6. If stimulation of the receptor is needed, remove the medium from each 60 mm dishes. Wash once with sterile PBS, aspirate, and add 2 mL of stimulation medium prepared in advance per dish. Add in the agonist at the desired concentration and stimulate for the desired times (*see* **Note 23**).

3.2.3 Cell Lysis

All samples should be kept on ice at all times to stop reactions and minimize protein degradation.

1. Remove the medium from each 60 mm dish. Wash once with 1 mL 1× cold PBS. Aspirate and add 1 mL 1× cold PBS to recover the cells.

2. Pellet the cells by centrifugation at $850 \times g$ for 2 min (*see* **Note 24**).

3. Discard the supernatant and resuspend the cells in 300 µL of cold lysis buffer (*see* **Note 25**). Incubate at 4 °C with rotation for 1 h.

4. Centrifuge the samples at $17,000 \times g$ for 15 min at 4 °C. Remove the gelatinous pellet that contains the insoluble material by pipetting it out of the tube.

5. To determine the protein concentration, use the Bio-Rad protein assay kit and follow the manufacturer's instruction. Adjust all samples to the same final protein concentration using lysis buffer.

6. Take a 20 µL sample from each cell lysate to use for SDS-PAGE analysis and add 20 µL of 4× SDS-PAGE sample buffer. Use 250 µL of the cell lysates for immunoprecipitation assays.

3.2.4 Immunopre-cipitation

1. Add 1 µg of FLAG M_2-specific monoclonal antibody to the 250 µL of cell lysates of each condition to immunoprecipitate the receptor (*see* **Notes 26** and **27**; Fig. 2). Incubate for 1 h at 4 °C with rotation (*see* **Note 28**).

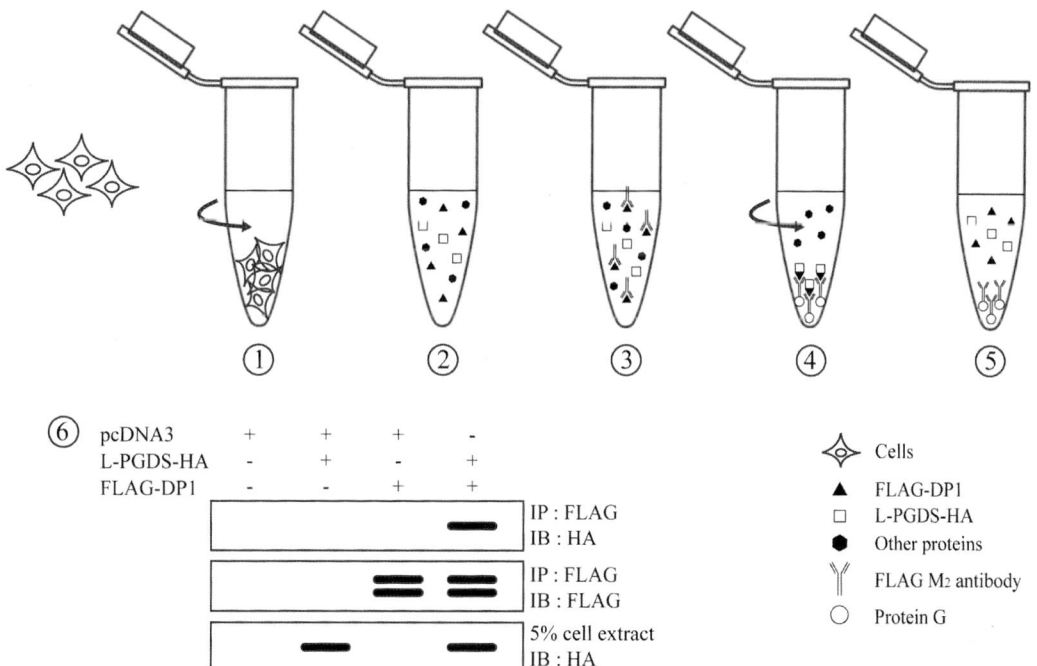

Fig. 2 Immunoprecipitation of a GPCR and analysis of interacting proteins by SDS-PAGE and western blotting. (*1*) Cells are collected and spun. (*2*) Cell pellets are resuspended in lysis buffer to release proteins. (*3*) The appropriate antibody is added, which then binds the protein of interest. (*4*) Protein G or A is added and binds the primary antibody used for immunoprecipitation. Supernatants are removed and antibody-protein complexes are washed. (*5*) Denaturation releases proteins from antibody/protein complexes. (*6*) Cell lysates and immunoprecipitation reactions are loaded onto SDS-PAGE

2. Add 40 µL of protein G-agarose beads to each tube and incubate for 30 min at 4 °C with rotation.

3. Centrifuge at 850×g for 2 min and discard the supernatants. Wash the beads three times with 1 mL cold lysis buffer.

4. Add 40 µL of 4× SDS-PAGE sample buffer to the samples, incubate for 1 h at room temperature (*see* **Note 29**), and proceed with SDS-PAGE analysis.

5. Prepare two SDS-PAGE gels. On the first gel, load 10 µL of the cell lysates. Load 20 µL of the immunoprecipitation reactions on gel two (*see* **Note 30**). After electrophoresis, transfer to nitrocellulose membranes (100 V, 60 min, in Minigel transfer cell).

6. Incubate membranes with blocking solution for 15 min at room temperature with gentle agitation.

7. Incubate the membranes with α-HA-HRP antibody (diluted 1:2,000 in blocking buffer) for 1 h at room temperature with gentle agitation (*see* **Note 31**).

8. Wash membranes three times for 8 min each with TBS-T.

9. Add ECL reagent, incubate, and expose to film.

10. After exposure to film, membranes can be stripped to remove the previous antibodies. To strip membranes, wash twice for 10 min with 0.2 M NaOH with gentle agitation and then wash twice for 7 min each with TBS-T with gentle agitation.

11. Block membranes with blocking solution for 15 min at room temperature with gentle agitation.

12. Incubate the membranes with α-FLAG polyclonal antibody (diluted 1:1,000 in blocking buffer) for 1 h at room temperature with gentle agitation.

13. Wash membranes three times for 8 min each with TBS-T.

14. Add α-rabbit secondary antibody to membranes (diluted 1:10,000 in blocking buffer) and incubate for 1 h at room temperature with gentle agitation.

15. Wash membranes three times for 8 min each with TBS-T.

16. Add ECL reagent, incubate, and expose to film.

3.3 G Protein-Coupled Receptor Immunoprecipitation (Endogenous Proteins)

Immunoprecipitations at the endogenous level are carried out following the same protocol as indicated in the immunoprecipitation section above with the following modifications.

3.3.1 Cell Lysis

All samples should be kept on ice at all times for best results.

1. Remove the medium from two 100 mm dishes seeded with HT29 cells at a 90 % confluence (*see* **Notes 32** and **33**). Wash once with 1 mL 1× cold PBS. Aspirate and add 1 mL 1× cold PBS to recover the cells.

2. Pellet the cells by centrifugation at $850 \times g$ for 2 min (*see* **Note 24**).

3. Discard the supernatant and resuspend the cells in 250 μL of cold lysis buffer (*see* **Note 25**). Incubate at 4 °C with rotation for 1 h.

4. Centrifuge the samples at $17,000 \times g$ for 15 min at 4 °C. Remove the gelatinous pellet that contains the insoluble material by pipetting it out of the tube.

5. Take a 20 μL sample from each cell lysate to use for SDS-PAGE analysis and add 20 μL of 4× SDS-PAGE sample buffer. Use the rest of the cell lysate for immunoprecipitations.

3.3.2 Immunoprecipitation

1. Add 5 μg of L-PGDS-specific monoclonal antibody to one condition to immunoprecipitate endogenous L-PGDS and HA-HRP antibody to the other condition for isotypic immunoprecipitation control (*see* **Notes 34** and **35**). Add 40 μL of protein G-agarose beads to each tube and incubate the tube overnight at 4 °C with rotation.

2. Centrifuge at $850 \times g$ for 2 min and discard the supernatants. Wash the beads three times with 1 mL cold lysis buffer.

3. Add 40 μL of 4× SDS-PAGE sample buffer to the samples, incubate for 1 h at room temperature (*see* **Note 29**), and proceed with SDS-PAGE analysis.

4. Prepare two SDS-PAGE gels (10 % acrylamide). On the first gel, load 10 μL of the cell lysates. Load 25 μL of the immunoprecipitation reactions on gel two. After electrophoresis, transfer to nitrocellulose membranes.

5. Proceed with immunoblotting as described above. Always immunoblot for the endogenous co-immunoprecipitated GPCR first (*see* **Note 31**).

4   Notes

1. The OverExpress strain C41 (DE3) was derived from BL21 (DE3). The OverExpress C41 strain is superior to the parental BL21(DE3) in transformation and expression of toxic proteins (OverExpress™ system Technical information).

2. GST pull-downs are carried out with the intracellular domains of the GPCR because we focus on intracellular interacting partners of the receptor. Moreover, GST-GPCR proteins are found in insoluble fractions and cannot be purified easily. It is recommended to study interaction with multiple intracellular domains even if interacting protein was isolated with a particular domain because GPCR partners can interact with more than one intracellular domain of the receptor.

3. The proper antibiotic to use is determined by the expression vector. In this case, since the constructs were subcloned into pGEX4T-1 or pRSETA expression vectors, ampicillin (diluted 1:1,000) was used because both vectors contain an ampicillin resistance gene.

4. For better results, allow cultures to reach an OD_{595} value of 0.6 for pRSETA constructs and of at least 0.8 for pGEX4T-1 constructs.

5. Induction may be carried out at different temperatures and for various amount of time depending on the efficacy of the purification.

6. To verify that protein expression has been induced, collect a 100 µL sample of the bacterial culture before and after induction, add 25 µL of 4× SDS-PAGE sample buffer, and run on an SDS-PAGE gel. The induced protein should appear blue after Coomassie staining.

7. Low levels of synthesis of the recombinant proteins may occur if the protein of interest has a tendency to accumulate in highly insoluble forms or into inclusion bodies. To prevent the formation of these insoluble forms and to enhance the level of purified proteins, sarkosyl detergent is added to the lysis buffer.

8. Glutathione Sepharose beads are supplied in a 20 % ethanol storage buffer. Remove the storage buffer by spinning the beads as indicated in the protocol and note the volume of storage buffer removed. After washing the beads to eliminate the ethanol present, replace the volume of storage buffer with 1× PBS containing 10 mg/mL BSA and resuspend the beads.

9. Intensities may vary according to the model of sonicator used. In this case, a BRANSON Sonifier 150 model was used.

10. To monitor the purity and verify that the protein of interest has not been lost at any particular step, a sample of each step of the purification process can be saved for analysis by SDS-PAGE.

11. Triton X-100 is a nonionic polyoxyethylene detergent frequently added in cell lysis buffers or other solutions to enhance the extraction and solubilization of proteins.

12. Adding glycerol to the binding buffer protects the purified proteins from damage during freezing and thawing.

13. It is recommended to freeze purified proteins in more than one aliquot to minimize the number of freeze/thaw cycles as these can denature the proteins.

14. Purified proteins can be used for up to 3 months if stored at −80 °C.

15. The purity of each GST-tagged protein may vary. Adjust quantities used for binding assay accordingly.

16. The binding assay can be carried out for 3 h or less if the His-tagged proteins tend to interact with beads alone.

17. Before loading, spin the binding reactions to pellet the beads and load the supernatants on the gels.

18. The secondary antibody varies according to the species used to generate the primary antibody.

19. In our studies, pcDNA3, pcDNA3-L-PGDS-HA, and pcDNA3-FLAG-DP1 served as immunoprecipitation controls [8]. Always prepare a separate set of controls for the unstimulated conditions and for each time point of the desired stimulation of the receptor.

20. Tagging the receptor at the N-terminus is recommended because many interacting partners bind the receptors at the C-terminal tail. An N-terminal tag is also convenient for detection by FACS and ELISA. Other tags can be used. Different tags work better for different GPCRs. Be aware of the presence of a signal peptide at N-terminal end of the receptor so as not to lose the tag during translation. It may be necessary to introduce a linker sequence between the tag and the receptor's N-terminus. It is our experience that incorporating multiple copies of the tag at the N-terminus of the receptor to increase detection is often detrimental to receptor trafficking and function. It is better to introduce a single copy of the tag.

21. It is recommended to optimize the amount of DNA used in transfection for best receptor expression and targeting. Too much DNA results in high GPCR expression but poor cell surface targeting due to ER-quality control overload, intracellular retention and ultimately to receptor degradation, and stress induction that affects cellular function. Better results in terms of expression and targeting of the GPCRs are often obtained by transfecting with smaller amounts of DNA.

22. To obtain the best results during transfection, at least 2 μg of total DNA should be employed for each 60 mm dish. If the quantities to add are less than 2 μg, complete to 2 μg of DNA with vector alone (in this case pcDNA3 since L-PGDS-HA and FLAG-DP1 were subcloned in this vector).

23. Here, a final concentration of 1 μM of PGD_2 was used to stimulate the DP1 receptor for 0, 5, and 30 min.

24. The cell pellets can be frozen at −80 °C before lysis. For better results, lysis and immunoprecipitation should be carried out on the same day and within a week after freezing.

25. To avoid sample dilution, use the smallest volume of lysis buffer possible. For a 60 mm dish seeded with HEK293 cells, a minimal volume of 250 μL of lysis buffer can be used.

26. Many antibodies can be used for immunoprecipitation according to the tag of the constructs used. Each DNA construct used should have different tags to facilitate immunoprecipitation and immunoblotting.

27. Immunoprecipitation should be carried out in both directions, that is, in our case, immunoprecipitation of FLAG-DP1 followed by immunoblotting of L-PGDS-HA, and immunoprecipitation of L-PGDS-HA followed by immunoblotting of FLAG-DP1.

28. Immunoprecipitations can be carried out overnight. In this case, antibody and bead are put in each condition at the same time. Overnight immunoprecipitation is recommended if the interaction studied is made between endogenous proteins.

29. Always allow the samples to denature at room temperature. Boiling a GPCR results in aggregation of the receptor and renders it impossible to detect by immunoblotting afterwards.

30. Before loading, spin the samples to pellet the protein G-agarose beads and load the supernatants on the gels.

31. Immunoblotting of the interaction partner should always be carried out first because the immunoblotting signal is lower than that of the immunoprecipitated protein and stripping a membrane can result in loss of protein and signal intensity.

32. The type of cells used for immunoprecipitating endogenous protein varies according to the endogenous expression of the proteins of interest. Not all cell types express the same proteins. Here HT29 cells were chosen because they express both endogenous L-PGDS and endogenous DP1 receptor [8].

33. If the endogenous protein of interest is difficult to see after immunoprecipitation and immunoblotting, a few cell dishes may be combined before lysis to increase the total amount of cells used.

34. Endogenous immunoprecipitation of the interacting partner is recommended because of the low levels of expression of endogenous GPCRs, the difficulty to immunoprecipitate GPCRs, and often the lack of good antibodies against GPCRs. Optimization of immunoprecipitation is necessary for each interacting protein as it is dependent on the cell type, the protein, and antibody used. Optimization of blotting conditions also may vary according to cell type, protein of interest, and antibody. Immunoblotting of GPCR should always be carried out first.

35. Isotypic control is determined according to the host antibody isotype used. Here, the L-PGDS monoclonal antibody was produced in the rat with an IgG_{1K} isotype. In this case, isotypic control should either be a rat IgG or an antibody produced with an IgG_{1K} isotype.

Acknowledgements

This work was supported by a grant from the Canadian Institutes of Health Research and by the André-Lussier Research Chair to J.L.P.

References

1. Luttrell LM (2008) Reviews in molecular biology and biotechnology: transmembrane signalling by G protein-coupled receptors. Mol Biotechnol 39:239–264

2. Conn PM, Ulloa-Aguirre A, Ito J et al (2007) G protein-coupled receptor trafficking in health and disease: lessons learned to prepare for therapeutic mutant rescue in vivo. Pharmacol Rev 59:225–250

3. Overington JP, Al-Lazikani B, Hopkins AL (2006) How many drug targets are there? Nat Rev Drug Discov 5:993–996

4. Tapaneeyakorn S, Goddard AD, Oates J et al (2011) Solution- and solid-state NMR studies of GPCRs and their ligands. Biochim Biophys Acta 1808:1462–1475

5. Maurice P, Guillaume JL, Benleulmi-Chaachoua A et al (2011) GPCR-interacting proteins, major players of GPCR function. Adv Pharmacol 62:349–380

6. Bockaert J, Perroy J, Bécamel C et al (2010) GPCR interacting proteins (GIPs) in the nervous system: roles in physiology and pathologies. Annu Rev Pharmacol Toxicol 50:89–109

7. Dupré DJ, Hébert TE (2006) Biosynthesis and trafficking of seven transmembrane receptor signaling complexes. Cell Signal 18:1549–1559

8. Binda C, Génier S, Cartier A et al (2013) A G protein-coupled receptor and the Intracellular synthase of its agonist functionally cooperate. J Cell Biol 204:377–393

9. Einarson MB, Pugacheva EN, Orlinick JR (2007) Preparation of GST fusion proteins. CSH Protoc. doi:10.1101/pdb.prot4738

Chapter 7

Isolation and Study of Cardiac Nuclei from Canine Myocardium and Adult Ventricular Myocytes

Artavazd Tadevosyan, Bruce G. Allen, and Stanley Nattel

Abstract

The nuclear envelope encloses the genome as well as the molecular machinery responsible for both the replication and transcription of DNA as well as the maturation of nascent RNA. Recent studies ascribe a growing number of functions to the nuclear membrane, in addition to sequestering the DNA, through receptors and their effectors, ion channels, as well as ion pumps and transporters located within the nuclear membrane itself. Despite the obvious structural and functional importance of the nucleus, certain aspects remain poorly understood due to the challenges associated with its accessibility in vivo, as well as isolating nuclei intact and with sufficient purity from cardiac cells to permit studies in vitro. Here, we present a detailed protocol for isolation of intact nuclei from both myocardial tissue and freshly isolated adult ventricular cardiomyocytes. These methods are based on partial permeabilization of plasma membrane with digitonin and cell disruption, followed by differential and discontinuous sucrose density centrifugation. These preparations provide for rapid separation of nonnuclear membranes and cytosol from nuclei.

Key words Cardiomyocyte, Myocardium, Intact nuclei, Nuclear GPCR, Subcellular fractionation

1 Introduction

The nucleus contains the genome as well as the molecular machinery responsible for both the replication and transcription of DNA as well as the maturation of nascent RNA. In contrast to most mammalian cells, nonpathological polynucleation is seen in cardiac myocytes as a result of karyokinesis during fetal growth [1]. Nuclei are enclosed within a double-lipid bilayer, the nuclear envelope (NE), which provides for selective entry of cytosolic proteins into the nucleus via the nuclear pore complex, but also serves as a platform for intracrine signaling processes that provide an additional level of control over gene expression. The nuclear envelope itself comprises the outer nuclear membrane (ONM) and inner nuclear membrane (INM), which are separated by a 40–50 nm perinuclear space. The perinuclear space is spanned by nuclear pore complexes (NPCs) and the inner surface of the INM is lined with a network

Bruce G. Allen and Terence E. Hébert (eds.), *Nuclear G-Protein Coupled Receptors*, Methods in Molecular Biology, vol. 1234, DOI 10.1007/978-1-4939-1755-6_7, © Springer Science+Business Media New York 2015

of nuclear intermediate filamentous structures formed by lamins. The ONM is contiguous with and contains many proteins in common with the peripheral endoplasmic reticulum. In contrast, the INM contains many distinct integral membrane proteins specific to the nucleus with direct interactions with lamina and chromatin [2, 3]. Recent experimental evidence demonstrates that many G protein-coupled receptors (GPCRs) including opioid, adrenergic, endothelin, and angiotensin receptors localize to the nuclear membrane and regulate gene expression [4–7], a signaling process often referred to as the "excitation-transcription coupling" [8, 9]. However, the molecular pharmacology and physiological function of receptors localized to the cardiac nuclear membranes remains poorly understood due in large part to the challenges associated with the accessibility of these receptors in vivo as well as in isolating intact nuclei that are viable and of sufficient purity from cardiac cells to permit studies in vitro.

In this chapter, we describe methods for isolating cytosolic, membrane, and nuclear fractions from canine ventricular myocardium and freshly isolated canine ventricular cardiomyocytes (Table 1). To isolate intact and enzymatically active nuclei, we selectively permeabilize the plasma membrane with digitonin, a cardiac glycoside and detergent that interacts with sterols. As nuclear membranes are not rich in cholesterol, their integrity is unaffected by digitonin treatment. Following permeabilization, nuclei are released from the remaining cellular organelles by gentle mechanical disruption using a Dounce homogenizer followed by differential sucrose density gradient centrifugation (Fig. 1) (*see* **Note 1**). The resulting nuclear-enriched fraction is suitable for functional studies.

Table 1
Size and density of cardiac subcellular organelles

	Diameter (μm)	Density in sucrose medium (g/cm³)
Nuclei	2–14	>1.40
Mitochondria	0.5–2	1.15–1.20
Golgi apparatus stacks	0.9–1	1.10–1.13
Peroxisomes	0.2–0.8	1.25
Lysosomes	0.2–0.8	1.2
Endoplasmic reticulum vesicles	0.05–0.30	1.15 (smooth ER) 1.25 (rough ER)
Plasma membrane vesicles	0.05	<1.15

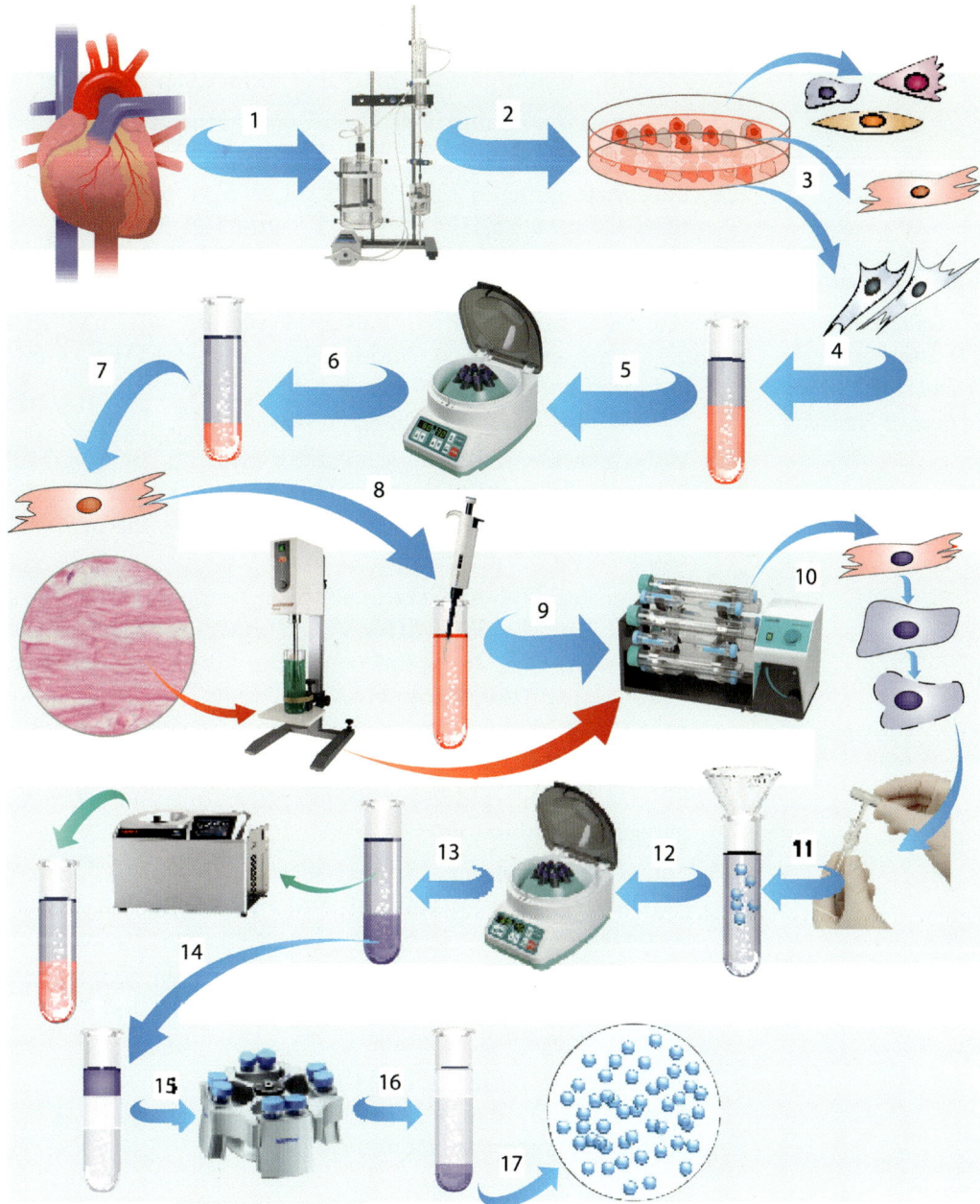

Fig. 1 Schematic outline of the subcellular fractionation

2   Materials

2.1   Solutions for Cardiomyocyte Isolation

1. Ultrapure (type 1) water must be used for all these procedures.

2. 2 mM Ca^{2+} Tyrode's solution: 136 mM NaCl, 5.4 mM KCl, 1 mM MgCl$_2 \cdot$ 6H$_2$O, 2 mM CaCl$_2$, 0.33 mM NaH$_2$PO$_4 \cdot$ H$_2$O, 5 mM HEPES, 10 mM glucose, pH 7.4 at room temperature.

3. Ca^{2+}-free Tyrode's solution: 136 mM NaCl, 5.4 mM KCl, 1 mM $MgCl_2 \cdot 6H_2O$, 0.33 mM $NaH_2PO_4 \cdot H_2O$, 5 mM HEPES, 10 mM glucose, pH 7.4 at room temperature.

4. Collagenase solution: 100 U/mL collagenase type II, 0.1 % bovine serum albumin (BSA) in Ca^{2+}-free Tyrode's solution.

5. Phosphate-buffered saline (PBS): 137 mM NaCl, 2.7 mM KCl, 4.2 mM $Na_2HPO_4 \cdot H_2O$, 1.8 mM KH_2PO_4, pH 7.4 at room temperature.

2.2 Solutions for Isolation of Membrane, Cytosol, and Nuclear Fractions

1. Homogenization buffer: 300 mM sucrose, 60 mM KCl, 0.15 mM spermine, 0.5 mM EGTA, 2 mM EDTA, 1 mM dithiothreitol (DTT), 10 mg/mL BSA, 1 mM $MgCl_2 \cdot 6H_2O$, 50 mM HEPES, pH 7.4 at room temperature.

2. 100× protease/phosphatase inhibitor cocktail: 20 mM NaF, 0.2 M Na_3VO_4, 20 mM β-glycerophosphate, 0.5 mM AEBSF, 25 μg/mL leupeptin, 10 μg/mL aprotinin, 1 μg/mL pepstatin, 1 μM microcystin.

3. 100× protease/phosphatase inhibitor and semi-permeabilization cocktail: 20 mM NaF, 0.2 M Na_3VO_4, 20 mM β-glycerophosphate, 0.5 mM AEBSF, 25 μg/mL leupeptin, 10 μg/mL aprotinin, 1 μg/mL pepstatin, 1 μM microcystin, 45 μg/mL digitonin.

4. Membrane storage buffer: 300 mM sucrose, 60 mM KCl, 0.5 mM EGTA, 2 mM EDTA, 1 mM DTT, 1 mM $MgCl_2 \cdot 6H_2O$, 50 mM HEPES, pH 7.4 at room temperature and protease/phosphatase inhibitors.

5. Nuclei storage buffer: 250 mM sucrose, 25 mM KCl, 5 mM $MgCl_2 \cdot 6H_2O$, 1 mM DTT, 50 mM HEPES, pH 7.4 at room temperature and protease/phosphatase inhibitor cocktail.

6. Low-sucrose buffer: 1.85 M sucrose, 25 mM KCl, 5 mM $MgCl_2 \cdot 6H_2O$, 1 mM DTT, 50 mM HEPES, pH 7.4 and protease/phosphatase inhibitors.

7. High-sucrose buffer: 2.15 M Sucrose, 25 mM KCl, 5 mM $MgCl_2 \cdot 6H_2O$, 1 mM DTT, 50 mM HEPES, pH 7.4 and protease/phosphatase inhibitors.

2.3 Equipment

1. Langendorff perfusion system.

2. Dounce homogenizer with both tight- and loose-fitting pestles.

3. Polytron PT3100 Homogenizer with a model PT-DA 3012/2 dispersing aggregate.

4. Refrigerated benchtop centrifuge with both fixed-angle and swinging-bucket rotors.

5. Beckman Avanti ultracentrifuge with swinging-bucket rotor.

6. Beckman Coulter Optima MAX Ultracentrifuge.

7. Polypropylene ultracentrifuge tubes (1.5, 15, 50 mL).

8. Confocal or inverted light microscope.

9. Western blotting system.

10. Nylon mesh (20, 60, 500 μm).

11. Liquid nitrogen.

12. Ice bucket, spatula, lab pipettes, funnel, digital balance.

13. Petri dish, microscope slides.

14. Sonicator, incubator.

15. 100 % O_2.

3   Methods

3.1   Canine Cardiomyocyte Isolation

3.1.1   Preparation

1. Turn on water baths. Pre-warm and oxygenate buffers (100 % O_2). Position the following close at hand: sets of sutures and surgical instruments.

2. Rinse the Langendorff perfusion apparatus with 75 % ethanol by recirculating for 15 min, and then rinse continuously with distilled H_2O followed by 2 mM Ca^{2+} Tyrode's perfusion solution. Throughout all the procedures carefully remove all air bubbles and preserve a constant flow rate (~12 mL/min).

3.1.2   Surgical Removal of the Heart

1. Anesthetize adult mongrel dogs of either sex (20–30 kg) with morphine (2 mg/kg subcutaneous injection), α-chloralose (120 mg/kg intravenous), and heparin (10,000 U) and mechanically ventilate (*see* **Note 2**).

2. Place the animal in a dissecting tray, shave the chest, and disinfect the area with alcohol.

3. Expose the heart through sternotomy by cutting and retracting the rib cage.

4. Open the pericardium, gently lift the heart at base with forceps, and support and excise heart out by cutting across the arch of the aorta and the vena cava, leaving a sufficient length of aorta.

5. Perform heart removal by a simple medical incision with the blunt end of a pair of blunt-sharp-pointed scissors to open the thoracic cavity.

6. Place the heart into 500 mL of ice-cold Ca^{2+} Tyrode's solution to rinse out blood. Remove all remaining fat, trachea, and connective tissue from the heart.

7. The perfusion of the entire heart is extremely costly. Hence, the left atria with a portion of the left ventricle are perfused. The heart is cut horizontally in the center to remove the lower section of the heart. The tissue that is removed is retained for subsequent subcellular fractionation.

3.1.3 Langendorff Perfusion and Digestion

1. Connect aortic sinus to the cannula of the perfusion system, secure with tie, and perfuse for 15 min with Ca^{2+} Tyrode's solution to rinse out any remaining blood from the heart. The tip of the cannula should be inserted above the base of the aorta to avoid ostial occlusion and restriction of the perfusion flow rate. Maintain a constant perfusate temperature of 36–38 °C (*see* **Note 3**).

2. Ligate any leaks from arterial branches with silk thread to assure adequate perfusate pressure and perfuse for 10 min with Ca^{2+}-free Tyrode's solution.

3. Perfuse the heart for approximately 60 min with 200 mL of collagenase solution by constantly recycling the solution (*see* **Note 4**).

4. Slice a small piece of tissue from a well-perfused region (soft and friable heart tissue), gently tear tissue apart with forceps, and triturate with a plastic transfer pipette to dissociate cell clumps.

5. Ensure that quiescent rod shape cardiac cells show clear tubular cross-striations under a light microscope (*see* **Note 5**).

6. Filter the cell suspension through a 500 µM nylon cell strainer into 50 mL tubes and centrifuge at 500 rpm ($55 \times g$) for 5 min with slow deceleration to pellet viable cardiac myocytes. Resuspend cells in 15 mL of ice-cold PBS, centrifuge at 500 rpm for 5 min, and repeat once more, to remove all the remaining fibroblast/debris (*see* **Note 6**).

3.2 Subcellular Fractionation

All the following steps should be performed on ice or at 4 °C.

3.2.1 Homogenization

For Myocardial Tissue

Dissect left ventricular tissue from fresh canine heart (~1 g), wash thoroughly in ice-cold PBS, and cut into small cubes/pieces (~2–3 mm²), with a scalpel/scissors. Transfer the tissue pieces into a 50 mL Falcon tube containing 15 mL of homogenization buffer. Homogenize the tissue with a Polytron PT-3100. Homogenizer, fitted with a model PT-DA 3012/2 dispersing aggregate, set at maximal speed (20,000 rpm) for 20 s, bringing the pestle to the bottom of the tube at least ten times.

For Cardiomyocytes

Following digestion, wash the myocytes with ice-cold PBS (*see* Subheading 3.1.3, **step 6**), centrifuge at 500 rpm ($55 \times g$) for 5 min at room temperature, and resuspend the cell pellet in 15 mL of homogenization buffer (*see* **Note 6**).

3.2.2 Differential Centrifugation

This step in the protocol applies to both tissue and cell homogenates.

1. Allow tissue/cells to permeabilize in homogenization buffer supplemented with protease/phosphatase inhibitors and

semi-permeabilization cocktail for 45 min under gentle agitation or rotation at 4 °C (*see* **Note 7**).

2. Dilute the homogenate with an equal volume of homogenization buffer supplemented with protease/phosphatase inhibitors (*without* digitonin). Transfer to a Dounce homogenizer. To further disrupt the tissue and cells and free the nuclei, perform 30 strokes using the loose-fitting "A" pestle followed by ten strokes with the tight "B" pestle. Sonicate for 20 s (*see* **Note 8**).

3. Centrifuge at 300 rpm ($20 \times g$) for 3 min at 4 °C (fixed-angle rotor, refrigerated benchtop centrifuge) to pellet and remove unbroken cells, connective tissue, and large contractile elements.

4. Filter the supernatant once through a 60 μm nylon mesh and then twice through a 20 μm nylon mesh. Additional filtering may be required to remove the bulk of the fibrous material and tissue remnants.

5. Transfer the filtrate to centrifuge tubes and pellet the isolated crude nuclei by centrifugation at 2,000 rpm ($850 \times g$) for 15 min at 4 °C in a swinging-bucket rotor (Sorvall 75-006-434 rotor).

6. Carefully decant off the supernatant into a clean centrifuge tube. Remove any remaining material from the inside of the tube with Kimwipes. Be careful not to disturb the soft pellet (nuclei).

7. Keep the supernatant. Gently resuspend the isolated crude nuclei in 5 mL of nuclei storage buffer. Ensure that the entire nuclear pellet at the bottom of the tube is collected. Evaluate the quality of isolated nuclei using a microscope.

8. Centrifuge the supernatant from **step 7** at 38,500 rpm ($80,000 \times g$) for 60 min at 4 °C (Beckman, TLA-100.3 rotor) to obtain the membrane (pellet) and cytosolic (supernatant) fractions (*see* **Note 9**).

9. Resuspend the pellet from **step 8** in membrane storage buffer and keep both the membrane and cytosolic fractions (supernatant) for further analysis.

3.2.3 Discontinuous Sucrose Density Gradient Centrifugation

1. Prepare a discontinuous sucrose gradient by pipetting 10 mL of freshly prepared high-sucrose buffer into a 30 mL ultracentrifuge tube (polyallomer, 25×89 mm). Carefully overlay the high-sucrose cushion with 10 mL of low-sucrose buffer. Finally, carefully layer the total resuspended crude nuclei pellet onto the discontinuous sucrose gradient (*see* **Note 10**). From this step forward, care must be taken not to disrupt the gradient.

2. Balance the ultracentrifuge tubes and centrifuge at 28,000 rpm ($141,000 \times g$) for 60 min at 4 °C in a swinging-bucket rotor (Beckman SW 28) (*see* **Note 11**).

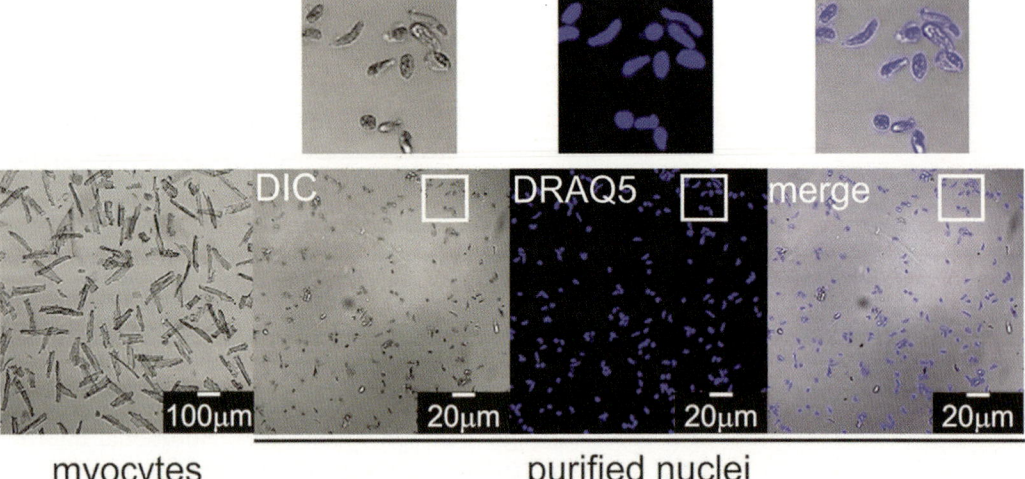

Fig. 2 Structural characterization of isolated cardiac cells and nuclei. *Left panel*: Isolated rod-shaped cardio-myocytes observed by differential interference contrast (DIC) microscopy. Remaining *lower panels*: Intact purified cardiomyocyte nuclei labeled with the fluorescent DNA dye DRAQ5 revealed by DIC or fluorescence microscopy. *Upper panels* contain enlargements of the field indicated by *white boxes* in the corresponding images in the *lower panel*

3. When the centrifugation is completed, carefully remove the tubes from the rotor and discard the supernatant (contains microsomal membranes and other cytoplasmic contaminants) using an aspirator. Wipe the remaining debris from the inside surface of the tube with Kimwipes (*see* **Note 12**).

4. Resuspend the pellet, containing pure nuclei, according to the needs of the experiment in nuclei storage buffer, transfer to a microcentrifuge tube, and snap freeze for later biochemical analysis. The isolated nuclei should be relatively free of contaminants and display an elongated (almond) or round shape and stain positive for DNA (Fig. 2). Further characterization of the isolated nuclei may be undertaken by immunofluorescence (Fig. 3) or immunoblotting (Fig. 4) (*see* **Note 13**).

4 Notes

1. Organelles may be purified according to their size and density by centrifugation (Table 1).

   Separation is determined by the centrifugal force employed. Centrifugal force may be determined for a particular centrifuge rotor using the following formula:

   $$\text{RCF}(g) = 1.12 \times r \times (\text{rpm} / 1{,}000)^2$$

   where RCF = relative centrifugal force (in $\times g$); r = maximum radius of the rotor (in mm); and rpm is the speed of rotation

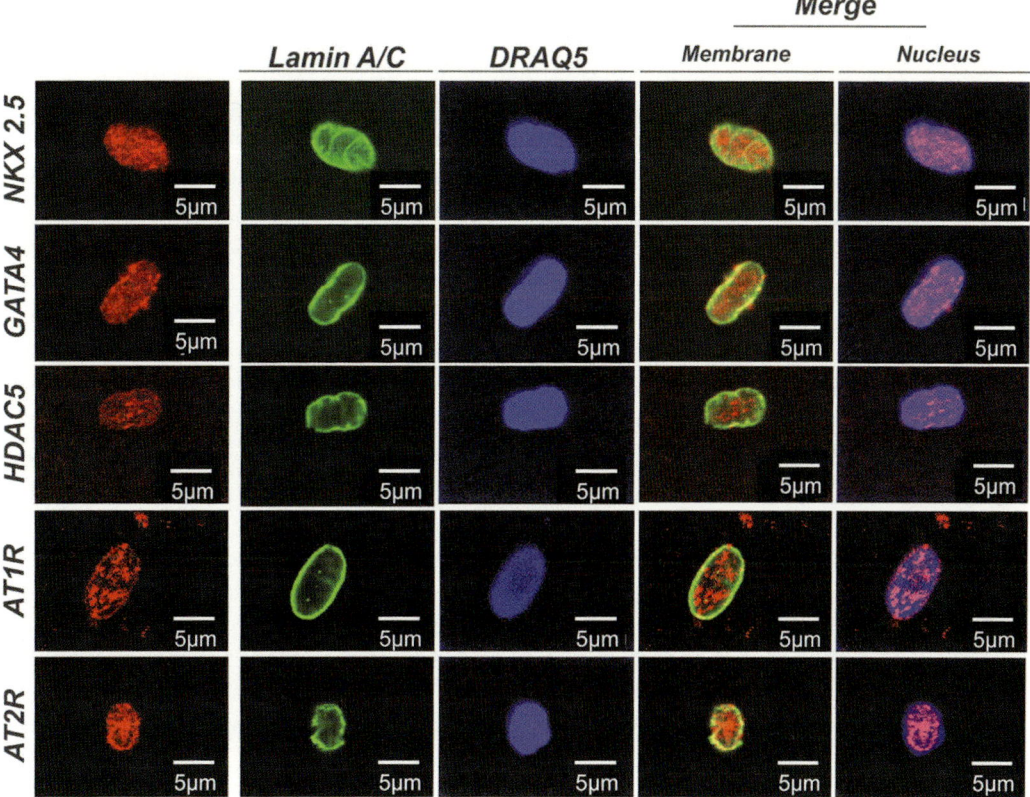

Fig. 3 Characterization of isolated canine cardiomyocyte nuclei by immunocytochemical staining. Cardiac nuclei were decorated with antibodies to Nkx2.5, GATA4, HDAC5, AT1R, and AT2R. Nuclei were also labeled with antibodies against lamin A/C and TO-PRO-3 to reveal the inner nuclear membrane and DNA, respectively

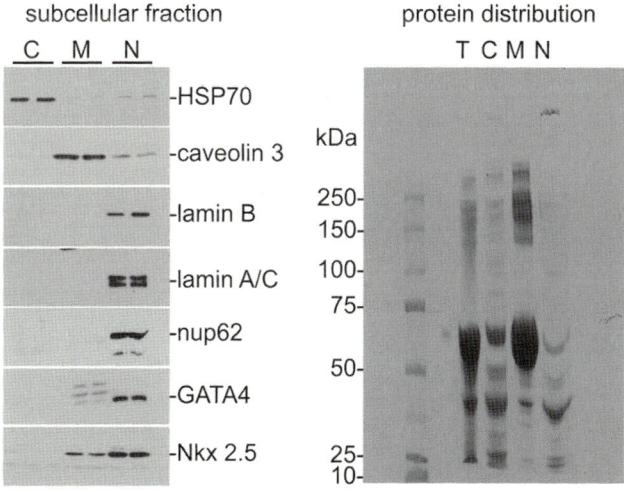

Fig. 4 Characterization of membrane, cytosolic, and nuclear fractions from canine cardiomyocytes. *Left panel*: Western blot analysis using markers for the cytoplasm (HSP70), plasma membrane (caveolin 3), inner nuclear membrane (lamins B and A/C), nuclear pore (nucleoporin 62), plus transcription factors that reside in cardiomyocyte nuclei (GATA4, Nkx2.5). *Right panel*: Fractions taken during nuclei isolation (membrane, cytosolic, and nuclear) were analyzed by SDS-PAGE (4–20 % gradient 12-well precast gel, 20 μg of total protein was loaded per lane) and visualized using Coomassie Brilliant Blue stain. T, total homogenate; C, cytosol; M, membrane fraction; N, nuclei

(in revolutions per minute). Therefore the RCF at the bottom of the tube is greater than at the top and denser organelles sediment faster than lighter particles. Moreover, the filling level and angle of the centrifugal tubes are also important factors to be considered. To take rotor angle into consideration, one must determine the pelleting efficiency or "k factor" for the rotor under the conditions employed.

2. Unless there is a specific experimental concern, animals should be heparinized prior to surgery to minimize blood clot formation in the vasculature and prevent formation of thrombi. Clotting, or the introduction of air bubbles during cannulation, will prevent uniform digestion of the myocardium by blocking access of perfusate to the tissue downstream of the blockage.

3. Care must be taken when positioning the aorta on the cannula to avoid puncturing the aorta/valve or trapping air bubbles. Once positioned on the cannula, the heart can be held in place with a blood vessel clamp (i.e., Dieffenbach serafine) to facilitate tying the heart onto the aortic cannula with sutures and to reduce time delays in initiating perfusion of the heart, as it only has oxygen and nutrients remaining in the vessels at the time of removal to sustain its metabolic needs. The processes of cannulation, attachment, and initiation of perfusion must be executed as rapidly as possible.

4. The time of digestion must be adjusted according to the canine heart size and pathophysiology to avoid under-digested cardiomyocytes with unbreakable membranes. Finding particular enzyme batches which dissociate cardiomyocytes without causing permanent damage is critical. Lots of collagenase must be tested until a suitable one is found. Moreover, when isolating cardiomyocytes from different species, one must modify digestion conditions accordingly (solution volumes, coronary perfusion rate, optimal enzyme perfusion time) to obtain viable cardiomyocytes in suitable number.

5. The isolation of rod-shaped Ca^{2+}-tolerant cardiomyocytes is normally reproducible and homogeneous populations of >95 % rod-shaped cells can be obtained routinely. However, if the viability decreases unexpectedly, this is a typical sign of bacterial growth, calcium contamination, or release of endotoxins into the perfusate. Initial preventive measures include (1) careful cleaning of the Langendorff apparatus and replacement of the Tygon tubing; (2) replacement of stock solutions as they have a limited storage life; and (3) inspection of distilled water source and pH of the oxygenated buffers at room temperature to ensure the quality of these solutions. The use of ultrapure (type 1) water is essential.

6. Nonviable cardiomyocytes are hypercontracted and hence pellet more slowly than viable cells. In our experience, survival of

cardiomyocytes often correlates with the quality of the isolation.

7. The osmotic strength of the homogenization buffer is crucial. A hypo-osmolar medium will cause organelles to swell and rupture. We therefore employ an iso-osmotic buffer containing 300 mM sucrose with EDTA and EGTA to chelate divalent metal ions that prevent organelle damage and protein degradation. The homogenization buffer contains spermine to stabilize chromatin.

8. Among the detergents (e.g., Triton-X 100, Tween 20, deoxycholate, digitonin) most commonly used to permeabilize membranes, digitonin is one that preferentially permeabilizes the plasma membrane of cardiac cells while leaving the nuclear envelope intact [10].

9. The size of the gap between the pestle and the glass Dounce homogenizer (pestel A: 0.0045 in, pestel B: 0.0025 in) is such that the rotating pestle applies a shearing force which breaks open the cell, releasing the contents, without breaking intracellular organelles. Therefore, if the pestle clearance is insufficient, or if too many strokes are applied, organelles will be damaged.

10. Alternatively, if the mitochondrial fraction is desired, centrifuge the initial supernatant (Sorvall, SS-34 rotor) at 10,000 rpm ($12,000 \times g$) for 15 min at 4 °C. This will pellet mitochondria, lysosomes, peroxisomes, and other organelles. Then use a continuous sucrose gradient (e.g., 15–35 %) to resolve these organelles.

11. Although sucrose is the most frequently used medium to generate density gradients, alternatives such as Ficoll, Percoll, Nycodens, or Metrizamide may be employed. However, the concentrations employed to prepare the discontinuous gradient will have to be optimized accordingly in order to permit isolation of nuclei.

12. Swinging-bucket rotors generally give a superior separation of cardiac nuclei from myofibrils, organelles, and cellular debris.

13. Nuclei from other tissues or species may have different densities; therefore the sucrose gradient as well as the conditions of centrifugation (e.g., time, RCF) may need to be optimized.

Acknowledgements

This work was supported by Operating Grants from the Canadian Institutes of Health Research and Quebec Heart and Stroke Foundation to SN. AT is recipient of FRSQ/RSCV-HSFQ doctoral scholarship.

References

1. MacLellan WR, Schneider MD (2000) Genetic dissection of cardiac growth control pathways. Annu Rev Physiol 62:289–319

2. Stewart CL, Roux KJ, Burke B (2007) Blurring the boundary: the nuclear envelope extends its reach. Science 318:1408–1412

3. Gerace L, Huber MD (2012) Nuclear lamina at the crossroads of the cytoplasm and nucleus. J Struct Biol 177(24):31

4. Tadevosyan A, Vaniotis G, Allen BG et al (2012) G protein-coupled receptor signalling in the cardiac nuclear membrane: evidence and possible roles in physiological and pathophysiological function. J Physiol 590: 1313–1330

5. Kang J, Shi Y, Xiang B et al (2005) A nuclear function of beta-arrestin1 in GPCR signaling: regulation of histone acetylation and gene transcription. Cell 123:833–847

6. Tadevosyan A, Maguy A, Villeneuve LR et al (2010) Nuclear-delimited angiotensin receptor-mediated signaling regulates cardiomyocyte gene expression. J Biol Chem 285:22338–22349

7. Vaniotis G, Glazkova I, Merlen C et al (2013) Regulation of cardiac nitric oxide signaling by nuclear β-adrenergic and endothelin receptors. J Mol Cell Cardiol 62:58–68

8. Wu X, Zhang T, Bossuyt J et al (2006) Local $InsP_3$-dependent perinuclear Ca^{2+} signaling in cardiac myocyte excitation-transcription coupling. J Clin Invest 116:675–682

9. Merlen C, Farhat N, Luo X et al (2013) Intracrine endothelin signaling evokes IP3-dependent increases in nucleoplasmic Ca^{2+} in adult cardiac myocytes. J Mol Cell Cardiol 62:189–202

10. Jamur MC, Oliver C (2009) Permeabilization of cell membranes. Methods Mol Biol 588: 63–66

Chapter 8

High Resolution Imaging and Function of Nuclear G Protein-Coupled Receptors (GPCRs)

Vikrant K. Bhosle, Fernand Gobeil Jr., Jose Carlos Rivera, Alfredo Ribeiro-da-Silva, and Sylvain Chemtob

Abstract

The traditional view of G protein-coupled receptors (GPCRs) being inactivated upon their internalization has been repeatedly challenged in recent years. GPCRs, in addition to forming the largest family of cell surface receptors, can also be found on intracellular membranes such as nuclear membranes. Since the first experimental evidence of GPCRs at the nucleus in the early 1990s, approximately 30 different GPCRs have been localized at the nucleus by independent research groups, including ours. In this chapter, we describe several techniques commonly used for immuno-detection of nuclear GPCRs focusing on subcellular fractionation of proteins based on their localization and transmission electron microscopy (TEM) using primary cultured cells as well as tissue sections. We also describe the use of confocal microscopy to study nuclear calcium currents, which can further affect downstream events such as gene transcription, nuclear envelope breakdown, or its reconstruction and nucleocytoplasmic protein transport.

Key words Nucleus, G protein-coupled receptor, Protein fractionation, Electron microscopy, Nuclear calcium currents

1 Introduction

G protein-coupled receptors (GPCRs) constitute the largest family of cell surface receptors [1]. Although an oversimplification, heterotrimeric G proteins have been classically viewed as a link between GPCRs at plasma membrane (PM) and intracellular signaling events [2]. Rubins et al. [3] first identified G proteins in the nuclear envelop of isolated rat liver nuclei. It has been also reported that G proteins can translocate to the nucleus in response to the receptor stimulation by specific growth factors and can control cellular events such as mitosis [4].

The earliest evidence for a nuclear localization of a GPCR was provided by Lind and Cavanagh [5, 6] using radioligand binding

Bruce G. Allen and Terence E. Hébert (eds.), *Nuclear G-Protein Coupled Receptors*, Methods in Molecular Biology, vol. 1234, DOI 10.1007/978-1-4939-1755-6_8, © Springer Science+Business Media New York 2015

studies, which detected the presence of muscarinic cholinergic receptors in isolated nuclei from rabbit corneal and Chinese Hamster Ovary (CHO-K1) cells. Bhattacharya et al. [7] first reported the presence of functional prostaglandin E (EP) receptors in isolated nuclei from a variety of tissues. Different subtypes of EP receptors (EP_1, EP_2, and EP_4) were detected in nuclear membranes using multiple approaches from primary cells, including piglet brain microvascular endothelial cells (PBMEC). Later, Gobeil et al. [8] reported that stimulation of isolated nuclei expressing the aforementioned EP receptors by prostaglandin analogs resulted in pertussis toxin-sensitive, G protein-mediated nuclear calcium signals regulating specific transcription of genes such as endothelial nitric oxide synthase (eNOS). These results provided a plausible mechanism for the contribution of prostaglandins (PGs), acting intracellularly, in local control of cerebral blood flow and oxygen delivery to neuronal tissues, critical during development [8].

Over the last decade, approximately 30 different GPCRs have been detected at nuclear membranes by independent research groups, including ours [9–11]. In some cases, the cellular machinery required for production of the GPCR agonist is already present at the nucleus. For example, enzymes such as cytosolic phospholipase A_2 ($cPLA_2$) [12], cyclooxygenase-1/2 (COX-1/2) [13], and some types of prostaglandin E synthases (PGE synthases) [13] have been detected at the nucleus. More and more GPCRs have been found to be functional at the nucleus via signaling mechanisms which are independent from those of cognate plasma membrane receptors [8, 14, 15]. More recently, receptors of tyrosine kinase (RTK) family (e.g., vascular endothelial growth factor receptor type-2; VEGFR2) have also been shown to translocate to the nucleus [16, 17]. VEGFR2 directly interacts with several transcription factors including Sp1 to regulate gene expression [16]. On the other hand, nuclear GPCRs like the type-1 lysophosphatidic acid receptor (LPA_1) [18] and β-adrenergic receptor (βAR) [19] can regulate gene expression by modulating cardiovascular nitric oxide signaling. Possible mechanisms for delivery of GPCRs and other receptors to the nuclear membrane are shown in Fig. 1.

Diverse types of primary cells (e.g., hepatocytes, cardiomyocytes, vascular endothelial cells), cancer cell lines, and stably transfected cell lines (e.g., human embryonic kidney cells) have been used to demonstrate nuclear localization of GPCRs. Common experimental approaches to localize receptors include protein fractionation according to subcellular localization, transmission electron microscopy (TEM) using primary cultured cells as well as using tissue sections and confocal microscopy [7, 8, 15, 20]. The latter can be utilized to confirm nuclear localization as well as to study nuclear calcium signaling. A brief rationale for each method is given below.

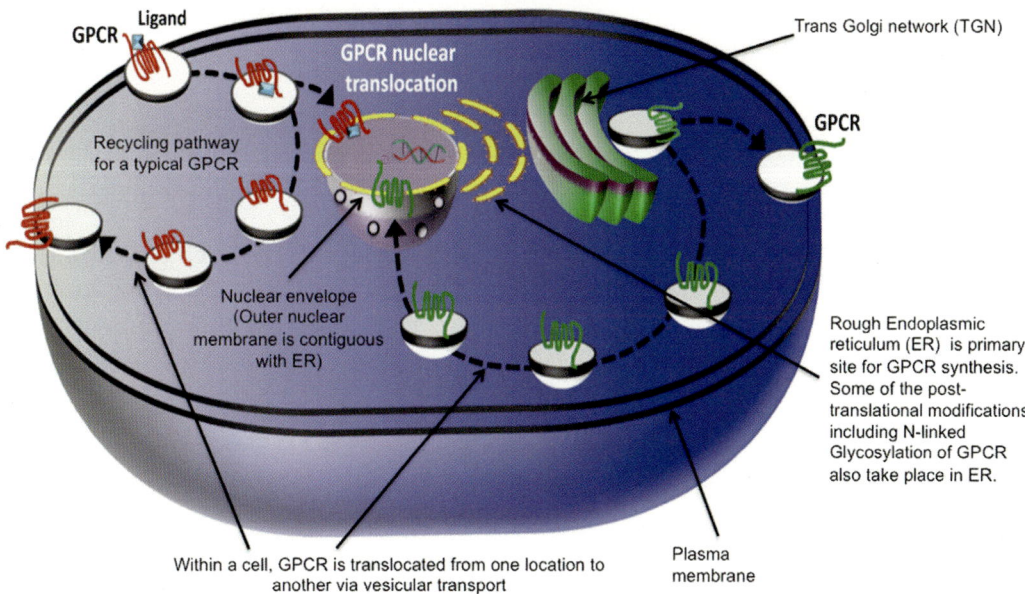

Fig. 1 Simplified schematic to show different pathways for nuclear translocation of a G protein-coupled receptor (GPCR). Little is currently known about mechanism of nuclear GPCR delivery. Different hypotheses are proposed [9–11]. Some propose that specific sequence motifs such as nuclear localization signal (NLS) might be responsible. In other cases, nuclear translocation might take place following agonist-induced internalization of the receptor from plasma membrane. The latter was recently reported to be the case [16] for a receptor tyrosine kinase (RTK) "vascular endothelial growth factor receptor type-2 (VEGFR2)." This mechanism might also be relevant for nuclear translocation of some, if not all, GPCRs

1.1 Subcellular Fractionation of Proteins from Cultured Cells to Study Nuclear Localization of GPCRs

Subcellular fractionation of cellular proteins, using either detergents [21] or ultracentrifugation [22], is commonly used to confirm their localization in different cellular compartments. However, isolation of nuclear membrane proteins without contamination from peri-nuclear organelles (e.g., endoplasmic reticulum, mitochondria) still remains a practical challenge [23]. We describe a simple protocol to isolate nuclear fractions with reasonable purity (*see* **Note 1a, b**). The method described in Subheading 3.1 is based on the use of a hypotonic/Nonidet P-40 lysis buffer, which was originally described in ref. 24 and modified in refs. 8, 15 for CHO-K1 cells and piglet brain microvascular endothelial cells (PBMEC).

1.2 Using Transmission Electron Microscopy (TEM) to Confirm Nuclear Localization of a GPCR in Primary Cultured Cells

Subcellular fractionation is relatively simple but the approach is not without limitations. Impurities (from surrounding organelles) in pelleted fractions (including the nuclear fraction) are often present. The homogenization step (*see* Subheading 3.1, **step 6**) requires special care to preserve nuclear integrity without compromising purity.

Transmission electron microscopy provides a practical alternative to confirm nuclear localization for GPCRs of interest [7, 15, 18]. However, there is limited literature available on preparation of primary cell culture samples for TEM [25, 26]. The protocol described below (Subheading 3.2) is a modified version of the protocol originally described in ref. 18 using instructions supplied by manufacturer of the Nanogold® antibody conjugate (Nanoprobes Inc.) [27, 28].

1.3 Using TEM to Study the Nuclear Localization of a GPCR in Tissue Sections

The ultrastructural demonstration of nuclear GPCR localization in intact cells under normal physiological conditions is particularly important. We performed such a study to demonstrate nuclear localization of platelet-activating factor (PAF) receptor in nuclear membranes of microvascular endothelial cells and in porcine brain sections [15]. The protocol below (Subheading 3.3) represents an updated version of the protocol used in the aforementioned publication and can be adapted with minor modifications to other tissues and animal species. It has been used by one of us to study the nuclear localization of another GPCR [29]. The advantage of the pre-embedding silver-intensified immunogold protocol we propose is that it is very easy to reproduce.

1.4 Identification of Functions and Signaling of Nuclear GPCR in Cell-Free Nuclei

The functional relevance of GPCR immunoreactivity identified at nuclear sites can be evaluated in terms of modulatory role of nuclear GPCRs on nucleoplasmic calcium homeostasis and ensuing effects on gene transcription, as highlighted by several studies including ours [7, 8, 10, 18, 30]. The details for determination of gene transcripts elicited upon stimulation of nuclei with GPCR ligands, for which nuclei contain respective receptors, are presented in a sister chapter by *Vaniotis* et al. Ligand occupation of plasma membrane GPCRs, which are coupled mainly to heterotrimeric Gi and Gq proteins, results in activation of one or more members of phospholipase C (PLC) class of enzymes, and subsequently, in transient augmentation of inositol 1,4,5-trisphosphate (IP_3) and diacylglycerol (DAG) production. This is followed by transient elevations in the concentration of free intracellular calcium. All of these signal transducers also reside in the nucleus [9–11].

The protocol described in Subheading 3.4 is based on the use of laser scanning confocal microscopy to monitor spatiotemporal movements of nuclear Ca^{2+} within a single nucleus upon GPCR stimulation by diverse ligands including lipids (e.g., PGs, LPA) [8, 30], amino acids (e.g., glutamate) [14], and peptides (e.g., bradykinin) [31]. This protocol has been tested in isolated nuclei from rat hepatocytes.

2 Materials

2.1 Subcellular Fractionation of Proteins from Cultured Cells to Study Nuclear Localization of GPCRs

1. Culture medium for Chinese Hamster Ovary (CHO-K1) cells—Dulbecco's Modified Eagle Medium: Nutrient Mixture F-12 (DMEM/F12) with fetal bovine serum (FBS) (final concentration: 10 %) and penicillin-streptomycin (P/S) (final concentration: 100 I.U./mL penicillin and 100 µg/mL streptomycin) added.

2. Culture medium for piglet brain microvascular endothelial cells (PBMEC)—Endothelial Cell Medium (ECM) with endothelial cell growth supplement (ECGS) (ScienceCell™ Research Laboratories), FBS (final concentration: 5 %), and P/S (final concentration: 100 I.U./mL penicillin and 100 µg/mL streptomycin) added.

3. Lysis buffer—10 mM Trizma-HCl pH 7.4, 10 mM NaCl, 3 mM $MgCl_2$. Store at 4 °C for up to 1 month.

4. Complete, EDTA-free protease inhibitor cocktail tablets (Roche Diagnostics) (*see* **Note 1d**). Store at 4 °C.

5. Nonidet P-40 alternative (EMD biosciences, Inc). Store at room temperature.

6. Dounce tissue grinder set (volume: 7 mL).

7. Primary antibodies for immunoblot analysis in (Fig. 2).

   (a) Anti-calnexin antibody (Abcam®).

   (b) Anti-pan-cadherin antibody (EMD Millipore Corp).

   (c) Anti-lamin-B receptor antibody (Epitomics Inc., An Abcam® Company).

   (d) Anti-platelet activating factor receptor antibody (Cayman Chemical Company).

   (e) Anti-β-actin antibody (Santa Cruz Biotechnology Inc.).

2.2 Using Transmission Electron Microscopy to Confirm Nuclear GPCR Localization in Primary Cultured Cells

1. Lab-Tek® chamber slides (Thermo Fischer Scientific)—It is important to use slides chambers with plastic bottoms (Permanox®), not glass, for electron microscopy.

2. Culture medium for piglet brain microvascular endothelial cells (PBMEC)—Endothelial Cell Medium (ECM) with endothelial cell growth supplement (ECGS) (ScienceCell™ Research Laboratories), FBS (final concentration: 5 %), and P/S (final concentration: 100 I.U./mL penicillin and 100 µg/mL streptomycin) added.

3. The recipe to make 100 mM (0.1 M) sodium phosphate buffer (PB) [32]:
   Add 3.1 g of $NaH_2PO_4 \cdot H_2O$ and 10.9 g of Na_2HPO_4 (anhydrous) to distilled H_2O to make final volume of 1 L. The pH of the final solution will be 7.4 at room temperature.

4. Fixation Buffer for first fixation (Subheading 3.2, **step 4**)—4 % paraformaldehyde, 0.25 % glutaraldehyde, 50 mM sucrose, 0.4 mM $CaCl_2$ in 0.1 M sodium phosphate buffer (pH 7.4).

5. Use heat shock-treated bovine serum albumin (fraction V) (Thermo Fischer Scientific) to prepare all solutions containing bovine serum albumin (BSA).

6. Washing Buffer (Subheading 3.2, **steps 14**, **15**, and **18**)—To prepare the washing buffer, mix 100 mL of 0.01 M PBS with 0.5 g fish gelatin and 0.8 g of BSA and adjust final pH to 7.4.

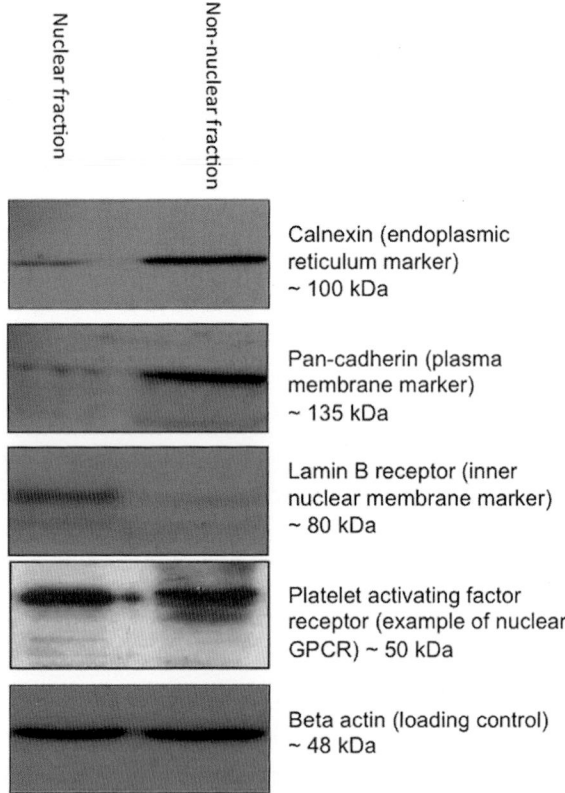

Fig. 2 Purity of nuclear and non-nuclear fractions was studied in subcellular fractions of piglet brain microvascular endothelial cells (PBMECs) analyzed for organelle-specific protein markers. Both nuclear and non-nuclear fractions were subjected to SDS-PAGE (9 % acrylamide) followed by immunoblot analysis. *Lane-1* corresponds to the nuclear fraction and *lane-2* to the non-nuclear fraction, which includes plasma membrane and cytosolic proteins. (1) Calnexin is a marker associated with endoplasmic reticulum (ER) membrane. (2) Cadherins are single chain glycoprotein receptors expressed at plasma membrane in tissue-specific manner. (3) Lamin B receptor (LBR) is an integral membrane protein associated with inner nuclear membrane. (4) Platelet activating factor receptor (PAFR) is an example of a GPCR found at the plasma membrane as well as at the nuclear membrane in piglet brain microvascular endothelial cells (PBMEC) [15]. (5) β-actin was used as loading control since it is found in both cytoplasm as well as nucleoplasm [33]

7. The secondary antibody used in this protocol is the Nanogold®-Fab' (Nanoprobes Inc.) fragment of goat anti-rabbit (host for primary antibody) IgG. The final Fab' concentration is 80 μg/mL. Store Nanogold®-Fab' at 4 °C (do not freeze). The concentrations provided for primary and secondary antibodies are for reference only. Final concentrations must be determined for each experiment.

8. Nanoprobes HQ SILVER™ kit (Nanoprobes Inc.) for silver intensification [28]. Store at –20 °C.

9. Use a digital camera and imaging software to acquire the images.

2.3 Using TEM to Study the Nuclear GPCR Localization in Tissue Sections

1. Perfusion mixture (Subheading 3.3, **step 1**)—3 % paraformaldehyde, 0.1 % glutaraldehyde, and 15 % picric acid (v/v) in 0.1 M sodium phosphate buffer (PB) pH 7.4.

2. Use heat shock-treated bovine serum albumin (fraction V) (Thermo Fischer Scientific) to prepare all solutions containing bovine serum albumin (BSA).

3. Washing Buffer (Subheading 3.3, **steps 9–11**)—To prepare the washing buffer, mix 100 mL of 0.01 M PBS with 0.5 g of fish gelatin and 0.8 g of BSA and adjust final pH to 7.4.

4. The secondary antibody used in this protocol is Nanogold®-Fab' (Nanoprobes Inc.) fragment of goat anti-rabbit (host for primary antibody) IgG. The final Fab' concentration is 80 µg/mL. Store Fab'-Nanogold® at 4 °C (do not freeze). The concentrations provided for primary and secondary antibodies are for reference. Final concentrations must be determined for each experiment.

5. Nanoprobes HQ SILVER™ kit (Nanoprobes Inc.) for silver intensification [28]. Store at –20 °C.

6. Use a digital camera and imaging software to acquire the images.

2.4 Identification of Functions and Signaling of Nuclear GPCR in Cell-Free Nuclei

1. The fluorescent calcium indicator Fluo-4AM (Molecular Probes; now part of Life Technologies) and fluorescent green nucleic acid dye SYTO-11 (Molecular Probes).

2. Glass coverslips for confocal fluorescence microscopy (Circles No. 1, 25 mm diameter, 0.13–0.17 mm thickness, Thermo Fischer Scientific).

3. Incubation buffer—25 mM HEPES pH 7.2, 125 mM KCl, 4 mM $MgCl_2$, 2 mM K_2HPO_4, 400 nM $CaCl_2$, and 0.5 mM ATP.

3 Methods

3.1 Subcellular Fractionation of Proteins from Cultured Cells to Study Nuclear Localization of GPCRs

1. Cell Culture—Seed cells of interest at initial density of 100,000 cells per 10 cm plate (*see* **Note 1c**). Grow them to an approximately 80 % confluence. The protocol below uses $\approx 50 \times 10^6$ cells as a starting material. Starve the cells for 4 h in the appropriate basal medium lacking in growth supplements and fetal bovine serum (FBS).

2. All the following steps are carried on ice unless mentioned otherwise.

3. Wash cells with ice-cold PBS and gently scrape them into 15 mL tubes.

4. Pellet the cells at $500 \times g$ for 5 min at 4 °C.

5. Resuspend the cell pellet in 2 mL Lysis buffer with complete (EDTA-free) protease inhibitors (*see* **Note 1d**).

6. Homogenize the suspension with approximately 100 gentle strokes (or for 20 min) with Dounce tissue grinder on ice (*see* **Note 1e**).

7. Centrifuge the homogenized suspension from Subheading 3.1, **step 6** at $600 \times g$ for 10 min at 4 °C. Store the supernatant (non-nuclear fraction) at –20 °C or refer to **Note 1f** for further processing.

8. Resuspend the nuclear pellet from Subheading 3.1, **step 7** in 2 mL Lysis buffer containing 0.1 % (v/v) Nonidet P-40 Alternative (*see* **Note 1g**) and protease inhibitors.

9. Leave suspension from Subheading 3.1, **step 8** on ice for 10 min and sediment the pellet at $600 \times g$ for 10 min at 4 °C.

10. Wash nuclear pellet with 10 mL lysis buffer containing 0.1 % (v/v) NP-40 Alternative and protease inhibitors for three times.

11. Leave suspension from Subheading 3.1, **step 10** on ice for 10 min, followed by sedimentation of nuclear pellet at $600 \times g$ for 10 min at 4 °C after each wash.

12. At the end of the washes, resuspend final nuclear pellet in 500 µL of lysis buffer with 0.1 % (v/v) Nonidet P-40 Alternative and protease inhibitors.

13. Store both protein fractions (nuclear and non-nuclear) at –80 °C for long term.

3.2 Using Transmission Electron Microscopy to Confirm Nuclear GPCR Localization in Primary Cultured Cells

1. Grow cells in Lab-Tek® chamber slides until they are about 80 % confluent (*see* **Note 2a**).

2. Wash cells twice with pre-warmed PBS (5 min per wash).

3. Following steps are done on a shaker (low speed).

4. First Fixation—Fix cells at room temperature (RT) for 30 min in fixation buffer (*see* **Note 2b, c**).

5. Wash cells twice with 1 % BSA in PBS (5 min per wash).

6. Treat with 1 % sodium borohydride in PBS for 30 min (*see* **Note 2d**).

7. Wash with PBS until bubbles completely disappear (usually four or five 10 min washes).

8. Block for 1 h with 1 % BSA in PBS.

9. Wash with PBS twice for 5 min each.

10. Permeabilize the cells for 15 min with 0.2 % Triton X-100 and 1 % BSA in PBS at RT.

11. Prepare primary antibody dilutions (10× more concentrated than used for confocal microscopy) in 1 % BSA in PBS.

12. Incubate primary antibody overnight at 4 °C in a humid chamber (*see* **Note 2e**). Do not forget negative control (no primary antibody) for each condition.

13. Wash with PBS three times for 10 min each.

14. Block for 15 min with Washing buffer.

15. Prepare gold-conjugated IgG dilution (1:500 as starting dilution) in Washing buffer.

16. Incubate chamber slides with the secondary antibody overnight at 4 °C in humid chamber. Avoid contact of chamber slides with any metal following secondary antibody incubation, as it will interfere with silver intensification (Subheading 3.2, **step 22**).

17. Remove Silver intensification kit (Nanoprobes HQ Silver™) from the freezer. It has to be at room temperature before use.

18. Wash chamber slides with Washing buffer for three times (1 min per wash).

19. Second Fixation—Fix cells at RT for 10 min in 1 % glutaraldehyde in PBS.

20. Wash with distilled water (dH$_2$O) twice (1 min per wash).

21. Silver Intensification (with Nanoprobes HQ Silver™): combine reagents A (initiator) + B (moderator) (mix well, viscous) then add C (activator) at a ratio of 1:1:1 just prior to use (*see* **Note 2f**).

22. Put 500 μL of reagent mixture into each chamber in the dark room.

23. Incubate for 8 min. Lightly shake the chamber slides during the incubation.

24. As the solution is very viscous, at the end of 8 min, try to take some solution out and then flush with dH$_2$O. Remove all the water immediately.

25. Wash twice with dH$_2$O for 5 min each.

26. Wash once with 0.1 M phosphate buffer, pH 7.4 for 5 min.

27. Incubate for 1 h in 1 % osmium tetroxide in 0.1 M phosphate buffer.

28. To dehydrate the cells:

    10 min wash with 30 % ethanol.

    10 min wash with 50 % ethanol.

    10 min wash with 70 % ethanol.

10 min wash with 90 % ethanol.

10 min wash with 95 % ethanol.

10 min wash with 100 % ethanol (three times).

29. Place in Epon: 100 % ethanol mixture in 1:1 ratio for 1 h.

30. Place in Epon: 100 % ethanol mixture in 2:1 ratio for 1 h.

31. Place in Epon: 100 % ethanol mixture in 3:1 ratio for 1 h.

32. Embed in pure Epon for 1 h.

33. Cure in oven at 60 °C for 48 h.

34. Remove the embedded cells from the Lab-Tek® chamber slide by peeling off the slide and then breaking off the sides of the chamber.

35. Obtain ultrathin sections in an ultramicrotome equipped with a diamond knife. Collect them on single-slot Formvar coated grids, contrast stain them with 4 % uranyl acetate in water and Reynold's lead citrate, and observe under the TEM.

3.3 Using TEM to Study the Nuclear Localization of a GPCR in Tissue Sections

1. Animals are anesthetized with sodium pentobarbital (100 mg/kg) and then perfused through the left ventricle briefly with saline followed by with the perfusion mixture (refer to Subheading 2.3, **item 1** for preparation) for 30 min, followed by with the perfusion mixture without glutaraldehyde for further 30 min and, finally, with 10 % sucrose in 0.1 M phosphate buffer for 30 min as well (*see* **Note 3a**).

2. The tissue of interest (e.g., cerebral cortex) is removed from the animal. The region of interest trimmed with a knife in thick slices (or blocks) with a thickness of 5 mm or less and placed overnight in 30 % sucrose in 0.1 M phosphate buffer at 4 °C.

3. Freeze-thaw the tissue by direct immersion in liquid nitrogen for 30 s and quickly thawing it in 0.1 M phosphate buffer at room temperature. For this, the tissue is placed in a customized small ladle with tiny holes drilled in it.

4. Cut 50 μm thick sections using a vibratome and collect them into a tissue culture multi-well plate (24 wells, BD Falcon). The wells should be pre-filled with PBS (do not use Triton) (*see* **Note 3b**).

5. Replace PBS in the wells by a solution of 1 % sodium borohydride ($NaBH_4$) in PBS. After 30 min, wash extensively in PBS until all bubbling disappears (usually four to five 10 min washes).

6. Place sections for 30 min in PBS with 0.5 % BSA.

7. Incubate sections overnight at 4 °C in the primary antibody diluted in PBS with 0.1 % BSA, on a shaker (gentle shaking).

8. Wash sections 3×10 min each with PBS.

9. Block with washing buffer for 1 h.

10. Incubate sections overnight at 4 °C, with gentle shaking, in a gold-conjugated Goat anti-rabbit antibody, diluted 1:200 in Washing buffer (*see* **Note 3c**).

11. Wash sections thrice (1 min each) with Washing buffer, followed by three washes (1 min each) with PBS.

12. Place in each well 250 μL of 1 % glutaraldehyde for 10 min.

13. Wash once for 1 min in dH$_2$O.

14. Move sections to properly labeled glass scintillation vials containing dH$_2$O.

15. After a 1 min rinse, it is time to carry out the silver intensification. Use Nanoprobes HQ Silver™ kit: mix reagents A (initiator) + B (moderator) (mix well, viscous) then + C (activator) in a 1:1:1 ratio just prior to use (at the end of the glutaraldehyde incubation is a good time). Put 750–900 μL of mixture in each vial with 1 mL syringe (*caution*: this reagent is light sensitive, carry it in a room lit only by dim incandescent light). Lightly shake the vials to move sections around or put the vials in a rack on a nutating mixer covered with foil. Total incubation time: 8 min. As this solution is very viscous, at the end of the 8 min try to take some solution out, flush the vials with water, and remove it immediately.

16. Wash 2× (5 min each) with dH$_2$O and then for 5 min with 0.1 M PB.

17. Incubate for 1 h in 1 % osmium tetroxide in 0.1 M PB.

18. Start dehydration of the sections:

    5 min wash in 50 % ethanol.

    5 min wash in 70 % ethanol.

    5 min wash in 90 % ethanol.

    5 min wash in 95 % ethanol.

    2 × 10 min wash in 100 % ethanol.

    10 min wash in propylene oxide.

    2 h in Epon: propylene oxide mixture in 1:1 ratio.

    2 h in Epon: propylene oxide mixture in 2:1 ratio.

    Immerse in pure Epon for 2 h.

19. Flat embed sections in pure Epon on thick acetate foil taped to a plastic or metal plate, using plastic coverslips. Use wood applicator sticks to remove sections from vials and do not apply pressure on the coverslip to avoid damaging the tissue.

20. Cure sections in oven at 60 °C overnight.

21. Remove material from oven, label coverslips (while still attached to the acetate sheet) with a permanent pen on the

coverslip side (not on the Epon side), and detach the coverslips (with the Epon and sections attached) from the acetate foil using a razor blade. Store them in small cardboard boxes.

22. Select relevant fields for EM study with a light microscope. Photograph as needed.

23. Re-embed selected fields in Epon. For that, trim the plastic coverslip/Epon to the inside diameter the plastic covers on the base of plastic capsules whose tops have been cut out. Close the covers to hold the plastic coverslip/Epon containing the relevant field in place. For best results the plastic coverslip should not be removed and should face the plastic on the bottom of the capsule (*see* **Note 3d**).

24. Fill the capsule with Epon and cure at 60 °C for 48 h (do not forget to insert a label).

25. Trim the blocks and obtain ultrathin sections in an ultramicrotome equipped with a diamond knife, collect them on single-slot Formvar coated grids, contrast stain them with 4 % uranyl acetate in water and Reynold's lead citrate, and observe under the TEM.

3.4 Identification of Functions and Signaling of Nuclear GPCR in Cell-Free Nuclei

1. Load suspension of nuclei with Fluo-4AM (30 μM) for 45 min at room temperature. Wrap tubes in aluminum foil to protect from light.

2. Dilute the nuclear suspension with (20×) excess volume of incubation buffer (*see* Subheading 2.4, **item 3**).

3. Spin down at $700 \times g$ for 5 min at room temperature.

4. Discard the supernatant and resuspend in the same buffer.

5. Gently place 10 μL aliquot of nuclear suspension (about 250,000 nuclei/assay) into custom-made chamber (designed for imaging in aqueous solutions) containing 500 μL of incubation buffer.

6. Allow 5–10 min to sediment nuclei onto the glass coverslips forming the bottom of the chamber.

7. Analyze nucleoplasmic calcium signals in single isolated nuclei in a rapid scan mode by using a multi-probe laser scanning confocal system (Bio-Rad) equipped with an inverse phase epifluorescence microscope (Nikon Eclipse TE300) and a 60× Nikon Oil Plan achromat objective (*see* **Note 4a**).

8. Use an injection volume of 10 μL of agonists (or other agents) to limit fluid disturbance so as to avoid displacement of nuclei during the confocal analysis (*see* **Note 4b**).

9. Identification and delineation of the nucleus can be accomplished at the end of experiments with a nucleic acid fluorescent dye (i.e., Syto-11, 100 nM) (*see* **Note 4c, d**).

4   Notes

1. Subcellular Fractionation of Proteins from Cultured Cells to Study Nuclear Localization of GPCRs.

   (a) The purity of nuclear fraction can be confirmed by light microscopy using Trypan blue staining (>90 % intact nuclei) or by electron microscopy.

   (b) Western blotting results of nuclear and non-nuclear fractions using organelle-specific protein markers are shown in Fig. 2.

   (c) Use at least five to eight 10 cm plates per condition to obtain enough amount of nuclear protein.

   (d) One protease inhibitor complete cocktail tablet is used to prepare 50 mL of lysis buffer. Alternatively, to prepare aliquots (25×), dissolve one tablet in 2 mL of distilled water, aliquot, and store at −20 °C.

   (e) Alternatively, the suspension from Subheading 3.1, **step 5** can be homogenized using a 2 mL syringe and 23 G needle (with approximately 50 strokes) on ice.

   (f) The supernatant from Subheading 3.1, **step 6** is the non-nuclear fraction and can be subjected to ultracentrifugation to obtain mitochondrial and microsomal fractions as described in ref. 21.

   (g) Nonidet P-40 Alternative is a nonionic, non-denaturing detergent used for isolation of functional protein complexes from nuclei. The optimal concentration of Nonidet P-40 Alternative might need to be adjusted (range—0.1–0.5 % v/v) depending on cell type [8, 15, 24].

2. Using Transmission Electron Microscopy to Confirm Nuclear GPCR Localization in Primary Cultured Cells.

   (a) It is important to make sure cells are not more than 80 % confluent; otherwise they will detach during subsequent treatment.

   (b) First fixation step is critical to ensure cellular ultrastructure (including integrity of biomembranes) is maintained during immunolabeling and embedding.

   (c) Occasionally, we have used 4 % paraformaldehyde without glutaraldehyde, but the fixation quality is less satisfactory. In this case only, Subheading 3.2, **steps 6** and **7** can be omitted.

   (d) Sodium borohydride (NaBH$_4$) is used to quench the remaining free aldehyde groups following glutaraldehyde fixation. Ammonium chloride (50 mM) or glycine (50 mM) in PBS (pH 7.4) can be used in place of NaBH$_4$.

(e) Wrap paraffin film around Lab-Tek® chamber slides before placing them in humid chamber. A conventional slide box with wet brown papers placed inside can be used as humid chamber.

(f) The reagents used for silver intensification are light sensitive. Mix them in dark room with a small incandescent light just prior to use. At the end of second fixation is a good time.

3. Using Transmission Electron Microscopy to Study the Nuclear GPCR Localization in Tissue Sections.

(a) For better morphological preservation, animals can be perfused instead with a mixture of 4 % paraformaldehyde and 0.5 % glutaraldehyde in 0.1 M PB, followed by the same mixture without glutaraldehyde and finally with 10 % sucrose in 0.1 M PB (each perfusion step lasting for 30 min). However, some antibodies against GPCRs may not give sufficient signal with this second fixation protocol (Fig. 3).

(b) The number of sections placed in each well should not be excessive to prevent section overlap that will compromise exposure to reagents.

(c) Avoid contact of sections with metal following secondary antibody incubation, as it will interfere with silver intensification step.

(d) Do not re-embed with the plastic coverslip facing up because the specimen can detach easily from the block during pyramid trimming or ultramicrotomy

4. Identification of Function and Signaling of Nuclear GPCR in Cell-Free Nuclei.

(a) Keep image acquisition settings constant (e.g., laser line intensity, photometric gain, filter attenuation) throughout the experiment.

(b) Ensure that the baseline fluorescence signal is relatively stable prior to the stimulation of nuclei. This can be done by performing short scans. There should be no probe leakage during the experiment.

(c) Intensity of fluorescence of the calcium-Fluo-4 complex can be converted into absolute calcium concentration as described in ref. 34.

(d) Application of an ionomycin (20 μM) and calcium (1 mM) mixture to the chamber can be done as a positive control at the end of experiments to measure maximal fluorescence intensity attainable corresponding to the levels of active (de-esterified) form of the Fluo-4AM calcium indicator.

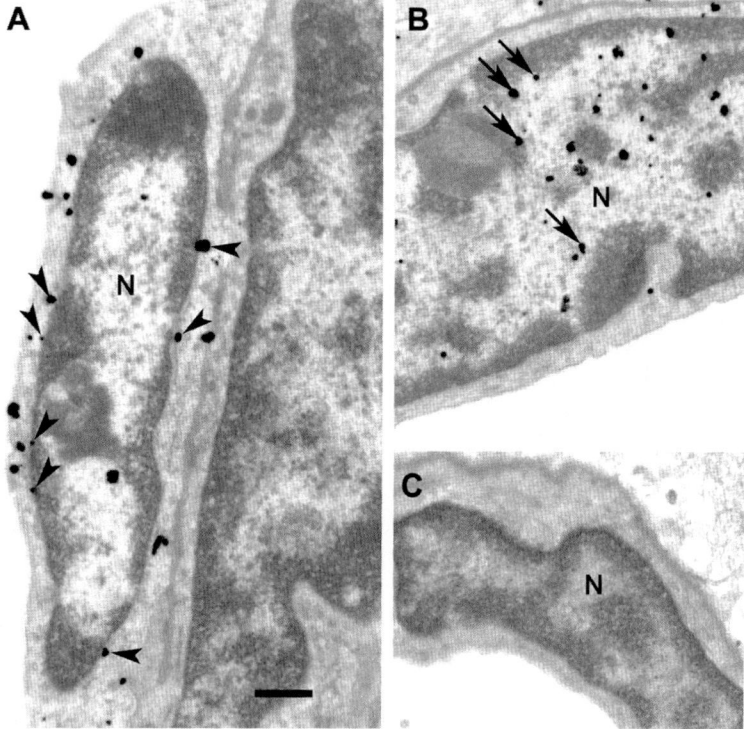

Fig. 3 Immunogold labeling for PAFR in microvascular endothelial cells from porcine cerebral cortex sections, obtained using a protocol similar to the one described in Subheading 3.3. In *panel* (**a**), note the localization of silver-gold grains over the nuclear membrane (*arrowheads*), which represent PAFR antigenic sites. In *panel* (**b**), note the predominantly intranuclear labeling in this endothelial cell (*arrows* indicate silver-gold grains). In *panel* (**c**), from material stained with the anti-PAFR antibody pre-adsorbed with the cognate peptide; note the complete absence of immunostaining. *N*-endothelial cell nucleus. Scale bar for all images = 0.5 μm

(e) Activation of mitogen-activated protein kinases (MAPK) by nuclear GPCR can be evaluated as previously described in refs. 18, 30. In brief, rat-liver derived nuclei (50 μg of proteins) are resuspended in aforementioned buffer (*see* Subheading 2.4, **item 3**) for calcium signal assay. In concomitant experiments, nuclear suspensions are pretreated with or without various pharmacological inhibitors of phosphatidylinositide 3-kinases (PI3K) prior to agonist-induced stimulation of GPCR of interest. Collected protein samples can be subjected to SDS-PAGE electrophoresis on a 9 % gel and immunoblotted for phospho-ERK1/2 (extracellular-signal-regulated kinases).

Acknowledgements

The electron micrographs shown in this chapter were obtained at electron microscopy facility in the Department of Pharmacology & Therapeutics, McGill University. The facility currently has a Philips CM120 electron microscope equipped with a Gatan digital camera. This work was supported by grants from Canadian Institutes of Health Research (CIHR). VKB is a recipient of CIHR Systems Biology studentship award at McGill University, Montreal, Canada. We thank Mrs. Hendrika Fernandez, Mrs. Isabelle Lahaie, and Ms. Johanne Ouellette for their technical expertise.

References

1. Gudermann T, Schoneberg T, Schultz G (1997) Functional and structural complexity of signal transduction via G-protein coupled receptors. Annu Rev Neurosci 20:399–427

2. Hamm H (1998) The many faces of G protein signaling. J Biol Chem 273:669–672

3. Rubins JB, Benditt JO, Dickey BF et al (1990) GTP binding proteins in rat liver nuclear envelopes. Proc Natl Acad Sci U S A 87:7080–7084

4. Crouch MF, Simson I (1997) The G-protein Gi regulates mitosis but not DNA synthesis in growth factor-activated fibroblasts: a role for the nuclear translocation. FASEB J 11:189–198

5. Lind GJ, Cavanagh HD (1993) Nuclear muscarinic acetylcholine receptors in corneal cells from rabbit. Invest Ophthalmol Vis Sci 34:2943–2952

6. Lind GJ, Cavanagh HD (1995) Identification and subcellular distribution of muscarinic acetylcholine receptor-related proteins in rabbit corneal and Chinese hamster ovary cells. Invest Ophthalmol Vis Sci 36:1492–1507

7. Bhattacharya M, Peri KG, Almazan G et al (1998) Nuclear localization of prostaglandin E2 receptors. Proc Natl Acad Sci U S A 95:15792–15797

8. Gobeil F Jr, Dumont I, Marrache AM et al (2002) Regulation of eNOS expression in brain endothelial cells by perinuclear EP(3) receptors. Circ Res 90:682–689

9. Zhu T, Gobeil F, Vazquez-Tello A et al (2006) Intracrine signaling through lipid mediators and their cognate nuclear G-protein-coupled receptors: a paradigm based on PGE2, PAF, and LPA1 receptors. Can J Physiol Pharmacol 84:377–391

10. Boivin B, Vaniotis G, Allen BG et al (2008) G protein-coupled receptors in and on the cell nucleus: a new signaling paradigm? J Recept Signal Transduct Res 28:15–28

11. Gobeil F, Fortier A, Zhu T et al (2006) G-protein-coupled receptors signalling at the cell nucleus: an emerging paradigm. Can J Physiol Pharmacol 84:287–297

12. Hirabayashi T, Shimizu T (2000) Localization and regulation of cytosolic phospholipase A2. Biochim Biophys Acta 1488:124–138

13. Vazquez-Tello A, Fan L, Hou X et al (2004) Intracellular-specific colocalization of prostaglandin E2 synthases and cyclooxygenases in the brain. Am J Physiol Regul Integr Comp Physiol 287:R1155–R1163

14. O'Malley KL, Jong YJ, Gonchar Y et al (2003) Activation of metabotropic glutamate receptor mGlu5 on nuclear membranes mediates intranuclear Ca2+ changes in heterologous cell types and neurons. J Biol Chem 278:28210–28219

15. Marrache AM, Gobeil F Jr, Bernier SG et al (2002) Proinflammatory gene induction by platelet-activating factor mediated via its cognate nuclear receptor. J Immunol 169:6474–6481

16. Domingues I, Rino J, Demmers JA et al (2011) VEGFR2 translocates to the nucleus to regulate its own transcription. PLoS One 6:e25668

17. Kojima Y, Nakayama M, Nishina T et al (2011) Importin β1 protein-mediated nuclear localization of death receptor 5 (DR5) limits DR5/tumor necrosis factor (TNF)-related apoptosis-inducing ligand (TRAIL)-induced cell death of human tumor cells. J Biol Chem 286:43383–43393

18. Gobeil F Jr, Zhu T, Brault S et al (2006) Nitric oxide signaling via nuclear endothelial nitric-oxide synthase modulates expression of the immediate early genes iNOS and mPGES-1. J Biol Chem 281:16058–16067

19. Vaniotis G, Glazkova I, Merlen C et al (2013) Regulation of cardiac nitric oxide signaling by nuclear β-adrenergic and endothelin receptors. J Mol Cell Cardiol 62:58–68

20. Chen R, Mukhin YV, Garnovskaya MN et al (2000) A functional angiotensin II receptor-GFP fusion protein: evidence for agonist-dependent nuclear translocation. Am J Physiol Renal Physiol 279:440–448

21. Holden P, Horton WA (2009) Crude subcellular fractionation of cultured mammalian cell lines. BMC Res Notes 2:43

22. Grataroli R, Termine E, Portugal H et al (1991) Subcellular localization of rat gastric phospholipase A2. Biochim Biophys Acta 1082: 130–135

23. Huber LA, Pfaller K, Vietor I (2003) Implications for subcellular fractionation in proteomics. Circ Res 92:962–968

24. Keller LR, Schloss JA, Silflow CD et al (1984) Transcription of alpha- and beta-tubulin genes in vitro in isolated Chlamydomonas reinhardi nuclei. J Cell Biol 98:1138–1143

25. Neumuller J (1997) Transmission and scanning electron microscope preparation of whole cultured cells. Methods Mol Biol 75:377–397

26. Heckman C (2008) Preparation of cultured cells for transmission electron microscope. Nat Protocol Exchange. doi: 10.1038/nprot. 2008.251. http://www.nature.com/protocol-exchange/protocols/518#/

27. Nanogold® manufacturer instructions. http://www.nanoprobes.com/instructions/Inf2001. html. Accessed 30 Nov 2013

28. Nanoprobes HQ SILVER™ instructions. http://www.nanoprobes.com/instructions/ Inf2012.html. Accessed 30 Nov 2013

29. Cornea-Hébert V., Ribeiro-da-Silva A., Coderre T.J. (2010) Nuclear localization of metabotropic glutamate receptor mGluR5, in spinal cord dorsal horn neurons and its possible role in neuropathic pain: an electron microscopic study. *Soc. Neurosci. Abstr. Program #175.13.*

30. Gobeil F Jr, Bernier SG, Vazquez-Tello A et al (2003) Modulation of pro-inflammatory gene expression by nuclear lysophosphatidic acid receptor type-1. J Biol Chem 278:38875–38883

31. Savard M, Barbaz D, Belanger S et al (2008) Expression of endogenous nuclear bradykinin B2 receptors mediating signaling in immediate early gene activation. J Cell Physiol 216: 234–244

32. Recipe: Sodium phosphate (2006) Cold Spring Harb Protoc. doi:10.1101/pdb.rec8303.

33. McDonald D, Carrero G, Andrin C et al (2006) Nucleoplasmic beta-actin exists in a dynamic equilibrium between low-mobility polymeric species and rapidly diffusing populations. J Cell Biol 172:541–552

34. Bkaily G, Jacques D, D'Orléans-juste P et al. (2001) In: Stanford C, Horton R (eds) Receptors, Oxford University Press, New York, p 209–232

Chapter 9

Biochemical Fractionation of Membrane Receptors in the Nucleus

Ying-Nai Wang, Longfei Huo, Jennifer L. Hsu, and Mien-Chie Hung

Abstract

Fractionation of cytoplasmic and nuclear proteins is a well-recognized biochemical technique to detect the intracellular distribution and expression level of proteins of interest. In the last decade, accumulating evidence shows that various types of cell surface receptors, such as receptor tyrosine kinases (RTKs), peptide hormone receptors, and cytokine receptors, are detected in the nuclei. Therefore, subcellular fractionation, including nonnuclear/nuclear extraction and the subsequent subnuclear fractionation without detectable cross-contamination during the process, is critical for studying membrane receptors that transit from the cell surface to the nucleus. Here, we utilize the epidermal growth factor receptor (EGFR) tyrosine kinase as an example of a comprehensive biochemical protocol for isolating membrane receptors in the nuclei of cancer cells.

Key words Nuclear extraction, Subnuclear fractionation, Sucrose gradient centrifugation, Inner nuclear membrane purification, Immunoblotting, Cell surface membrane receptor, Receptor tyrosine kinase, Epidermal growth factor receptor

---

1 Introduction

Emerging evidence suggests that mislocalization and aberrant compartmentalization of cellular proteins, such as receptor tyrosine kinases (RTKs), a major class of cell surface receptors consisting of 20 subfamilies that mediate extracellular signals in cells, are involved in the development of diverse diseases, including various types of cancer [1]. Since many RTKs are overexpressed and activated in human malignancies and frequently related to poor patient survival, a better understanding of the subcellular trafficking of RTKs will advance our knowledge of their unique functions at different cellular destinations and contribute to the clinical application of potential therapeutic targets of anti-RTK treatments [2]. Interestingly, 11 classes of RTKs to date, including subfamilies of ErbB, insulin, PDGF, VEGF, FGF, cMet/HGF, Trk, Ror, Mer, Eph, and Ryk receptors, can shuttle from the cell surface to the nucleus, where they carry out associated biological functions, such

Bruce G. Allen and Terence E. Hébert (eds.), *Nuclear G-Protein Coupled Receptors*, Methods in Molecular Biology, vol. 1234, DOI 10.1007/978-1-4939-1755-6_9, © Springer Science+Business Media New York 2015

as transcriptional regulation, DNA repair, DNA replication, and drug resistance. Other classes of cell surface receptors also translocate to the nucleus, such as peptide hormone (e.g., G protein-coupled receptors such as the oxytocin receptor, and the angiotensin II type 1 receptor) and cytokine (e.g., interferon-γ receptor and CD40 receptor) receptors [3–6]. Accordingly, the field of RTKs and cell surface receptors with noncanonical nuclear localization is termed as *m*embrane *r*eceptors *i*n the *n*ucleus (MRIN) [7].

Here, we utilize the epidermal growth factor receptor (EGFR) of the ErbB family, one of the best characterized RTKs in the MRIN field, to illustrate subcellular fractionation. In addition to the traditional trafficking route of cell surface EGFR to either the late endosomes/lysosomes for degradation or to the cell surface for recycling after endocytosis and endosomal sorting, a novel nuclear EGFR signaling pathway has been recently discovered [8–12]. EGFR is also found in the nuclear matrix or inner nuclear membrane (INM) of the nuclear membrane/envelope [13–15] where cell surface EGFR is targeted to the INM in response to EGF stimulation by a proposed mechanism named INTERNET, which stands for the *in*tegral *tr*afficking from the *e*ndoplasmic *r*eticulum (ER) to the *n*uclear *e*nvelope (NE) *t*ransport [16]. Therefore, to investigate important nuclear functions of cell surface membrane receptors that may have been overlooked in the past, subcellular fractionation, including nonnuclear/nuclear extraction to isolate the nuclei and the subsequent subnuclear fractionation to purify the NE composed of nuclear membranes, is a critical biochemical method for such MRIN studies. In this chapter, we provide a detailed fractionation procedure to demonstrate how cell surface membrane receptors can be isolated from the nuclei with minimum cross-contamination during the entire process.

2   Materials

Prepare all solutions using ultrapure water from Milli-Q water filtration station (18 MΩ cm at 25 °C) with analytical grade reagents, and store reagents at room temperature unless indicated otherwise.

2.1 Nonnuclear/ Nuclear Extraction Components

1. Phosphate-buffered saline (PBS): 137 mM NaCl, 2.7 mM KCl, 10 mM Na_2HPO_4, 1.8 mM KH_2PO_4. Mix and adjust pH with HCl to 7.4. Store at 4 °C.

2. Lysis buffer: 20 mM HEPES (pH 7.0), 10 mM KCl, 2 mM $MgCl_2$, 0.5 % Nonidet P-40 (v/v). Store at 4 °C. Add phosphatase and protease inhibitors including 1 mM Na_3VO_4, 100 mM NaF, 1 mM phenylmethanesulfonyl fluoride, and 2 % (v/v) aprotinin (*see* **Note 1**).

3. NETN buffer: 150 mM NaCl, 1 mM EDTA (pH 8.0), 20 mM Tris–HCl (pH 8.0), 0.5 % Nonidet P-40 (v/v) (*see* **Note 2**).

Store at 4 °C. Add phosphatase and protease inhibitors including 2 mM Na_3VO_4, 25 mM NaF, 1 mM phenylmethanesulfonyl fluoride, and 2 % (v/v) aprotinin (*see* **Note 1**).

4. Cell scrapers.

5. Dounce tissue grinder/homogenizer: 2 mL (Kontes Glass Co., Vineland, NJ, USA) and 7 mL (Wheaton, Millville, NJ, USA) (*see* **Note 3**).

6. Microcentrifuge tubes (1.5 mL) or centrifuge tubes (15 mL) (*see* **Note 4**).

7. Trypan blue stain solution: 0.4 % (w/v) solution in normal saline storage temperature 15–30 °C.

2.2 Subnuclear Fractionation Components

1. Buffer A: 0.25 M sucrose, 50 mM Tris–HCl (pH 7.4), 10 mM $MgCl_2$, 1 mM dithiothreitol. Store at 4 °C. Add protease inhibitor mixture (1:100; Sigma-Aldrich, St. Louis, MO, USA) (*see* **Note 1**).

2. Sodium citrate: 1 % (w/v) solution in Buffer A containing protease inhibitor mixture (*see* **Note 5**).

3. Deoxyribonuclease I (DNase I) (Sigma): 250 µg/ml final concentration dissolved in 0.15 M NaCl.

4. Sulfo-NHS-LC Biotin (Pierce, Rockford, IL, USA): 1 mM final concentration dissolved in 167 mL PBS (*see* **Note 6**).

5. Streptavidin, immobilized on Agarose CL-4B (Sigma): 50 % suspension in 0.01 M sodium phosphate buffer, pH 7.2, containing 0.15 M sodium chloride and 0.02 % sodium azide as preservative.

6. Quenching buffer (PBS containing 100 mM glycine): Dissolve 1.5 g glycine in 200 mL PBS.

2.3 Sucrose Gradient Centrifugation Components

1. Sucrose gradient: 0.25 M sucrose (dissolve 4.3 g sucrose in 50 mL water), 1.6 M sucrose (dissolve 27 g sucrose in 50 mL water), 2.4 M sucrose (dissolve 41 g sucrose in 50 mL water). Store all at 4 °C (*see* **Note 7**).

2. Ultra-clear thin-wall tubes (Beckman Coulter, Brea, CA, USA).

3. Optima XL-100 K Ultracentrifuge (Beckman Coulter).

4. SW 40 Ti Swinging bucket rotor (Beckman Coulter).

2.4 Chemicals, Antibodies, and Equipment

1. Epidermal growth factor (EGF): Dissolve one vial of EGF (0.2 mg; Sigma) to 100 µg/mL by adding 2 mL of 0.2 µm filtered 10 mM acetic acid. Store at −80 °C (*see* **Note 8**).

2. Antibodies: Anti-EGFR, anti-Lamin B, anti-calregulin, anti-calnexin, anti-Sp1, anti-CD44/HCAM, anti-emerin, and anti-LAMP1 (Santa Cruz Biotechnology, Dallas, TX, USA); anti-α tubulin (Sigma); anti-emerin Ab-1 (Thermo Scientific, West

Palm Beach, FL, USA) (*see* **Note 9**); anti-Rab5 (BD Biosciences, Franklin Lakes, NJ, USA).

3. Sonicators: Bioruptor sonication device (Diagenode, Denville, NJ, USA) and Ultrasonic Processor (Sonics Vibra-Cell, Sonics & Materials, Newtown, CT, USA).

4. Centrifuge 5415R (Eppendorf, Hauppauge, NY, USA).

3   Methods

3.1 Subcellular Nonnuclear/Nuclear Fractionation

1. Seed cells in an appropriate environment, e.g., A431 epidermoid carcinoma at a density of 2×10^6 per 10 cm culture dish in a 37 °C incubator containing 5 % CO_2.

2. Starve cells in serum-free medium overnight.

3. Stimulate cells with EGF (final concentration: 50 ng/mL) for 30 min (*see* **Note 10**).

4. Discard culture medium.

5. Wash cells with ice-cold PBS (pH 7.4) twice.

6. Add 800 μL of ice-cold lysis buffer containing phosphatase and protease inhibitors (*see* **Note 11**).

7. Scrape cells with a clean cell scraper.

8. Collect cells into a 1.5 mL pre-chilled microcentrifuge tube.

9. Incubate the cells on ice for 10 min (*see* **Note 12**).

10. Transfer half volume (about 400–500 μL) of cells into a pre-chilled 2 mL Dounce tissue homogenizer (*see* **Note 13**).

11. Homogenize cells gently with 25–30 strokes with a tight pestle on ice (*see* **Note 14**).

12. Transfer the homogenized cells into a new pre-chilled 1.5 mL microcentrifuge tube.

13. Homogenize another half volume of cells by repeating Subheading 3.1, **steps 10–12**.

14. Combine the homogenized cells together.

15. (*Optional*) Test cell viability and nucleus integrity by a trypan blue exclusion assay (*see* **Note 15**, Fig. 1).

16. Centrifuge the homogenized cells at $1,500 \times g$ for 5 min at 4 °C.

17. Transfer the supernatant carefully to a new 1.5 mL pre-chilled microcentrifuge tube without disturbing the nuclear pellet.

18. Centrifuge the supernatant for 20 min at $16,100 \times g$ at 4 °C.

19. Collect the supernatant after centrifugation as the nonnuclear/cytoplasmic fraction. Store at −80 °C.

20. Wash the nuclear pellet obtained from Subheading 3.1, **step 17**, with 1.0 mL lysis buffer by gently pipetting ten times (*see* **Note 16**).

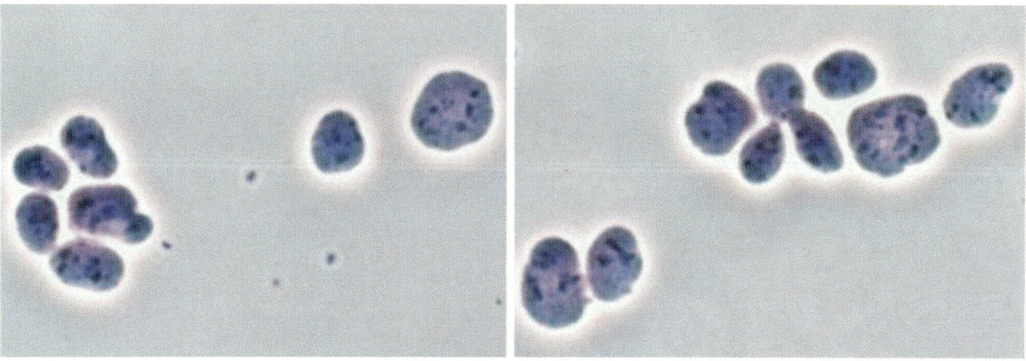

Fig. 1 Validation of cell viability and nucleus integrity using a trypan blue exclusion assay. Briefly, after homogenization, cells were washed five times with lysis buffer. Then, the homogenized cells were mixed with trypan blue (1:1, v/v) and incubated at room temperature for 2–3 min before examination under a microscope. Two representative images are shown

21. Centrifuge the suspended nuclear pellet at $1,500 \times g$ for 5 min at 4 °C.

22. Remove supernatant without disturbing the nuclear pellet.

23. Repeat the wash procedures in Subheading 3.1, **steps 20–22**, at least four times.

24. Resuspend the nuclear pellet in 100–120 µL of ice-cold NETN buffer containing phosphatase and protease inhibitors (*see* **Note 17**).

25. Sonicate the nuclear pellet using Bioruptor sonication device (*see* **Note 17**) or Ultrasonic Processor (*see* **Note 18**).

26. Centrifuge nuclear resuspension for 20 min at $16,100 \times g$ at 4 °C.

27. Collect the supernatant as the nuclear fraction. Store at –80 °C (*see* **Note 19**).

28. Perform immunoblotting by loading 50–100 µg/well of proteins (Fig. 2, *see* **Note 20**).

3.2 Subnuclear Inner and Outer Nuclear Membrane Purification

1. Take the nuclear pellet as intact nuclei extracted from Subheading 3.1, **step 23**, for subsequent subnuclear fractionation biochemically separated into various fractions, including the outer nuclear membrane (ONM), INM, and nucleoplasm (*see* **Note 21**, Fig. 3).

2. Suspend the nuclear pellet in Buffer A containing protease inhibitor mixture at a protein concentration of 2 mg/mL (*see* **Note 22**).

3. Incubate the nuclear suspension with 1 % (w/v) sodium citrate (*see* **Note 23**) and gently rotate it at 4 °C for 30 min.

4. Centrifuge for 15 min at $500 \times g$.

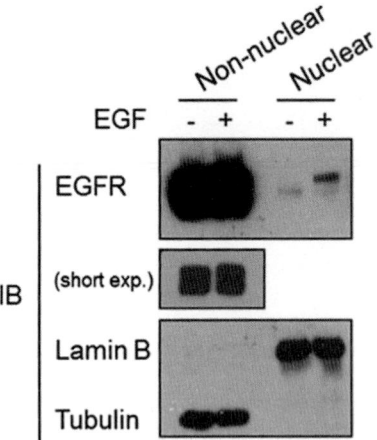

Fig. 2 Nuclear translocation of EGFR in response to EGF. A431 cells were treated with or without EGF (50 ng/mL) for 30 min and subjected to subcellular nonnuclear/nuclear fractionation. Proteins (80 μg) extracted from nonnuclear/cytoplasmic and nuclear fractions were subjected to immunoblotting (IB). Lamin B and tubulin were used as markers for nuclear fraction and nonnuclear fraction, respectively. Short exp., five times shorter

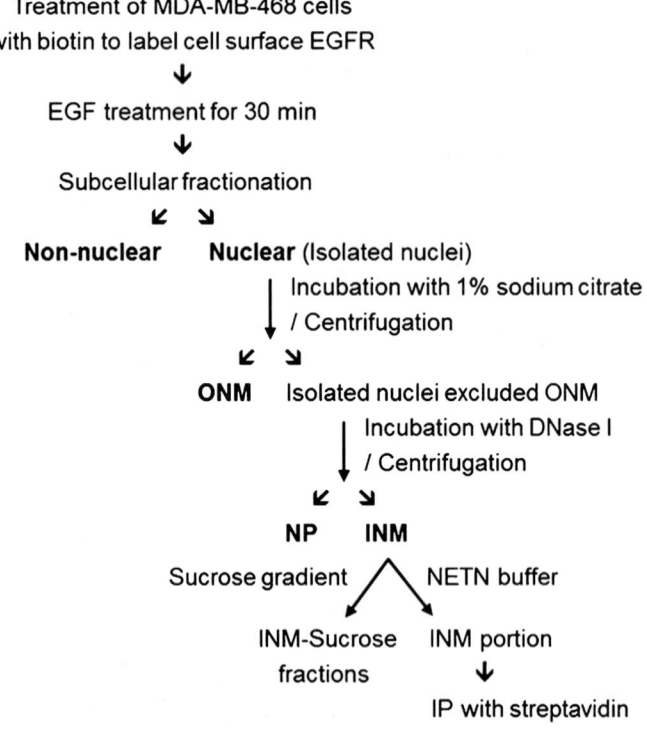

Fig. 3 A schematic description of subcellular fractionation of biotinylated cell surface proteins in MDA-MB-468 breast cancer cells. ONM, outer nuclear membrane; INM, inner nuclear membrane; NP, nucleoplasm; IP, immunoprecipitation. This research was originally published in ref. 13 and is reproduced with permission from the American Society for Biochemistry and Molecular Biology

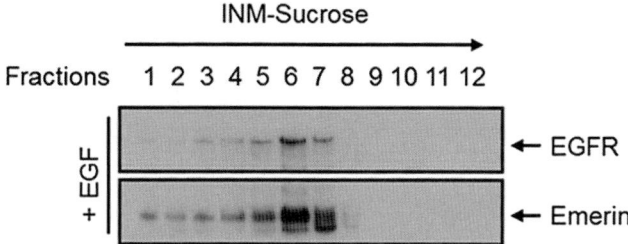

Fig. 4 Distribution of EGFR in the INM. INM-sucrose fractions were purified using sucrose gradient and subjected to immunoblotting with the indicated antibodies. The *arrow* above the panels indicates the direction of the gradient from *top* to *bottom*. This research was originally published in ref. 13 and is reproduced with permission from the American Society for Biochemistry and Molecular Biology

5. Collect the supernatant containing the ONM-enriched portion (*see* **Note 24**).

6. Suspend the remaining pellet that contains the isolated nuclei, excluding the ONM, with Buffer A at a protein concentration of 5 mg/mL (*see* **Note 22**).

7. Digest with DNase I (250 μg/ml) at 4 °C for 14–18 h (*see* **Note 25**).

8. Centrifuge the suspension for 2 h at $10,000 \times g$.

9. Collect the supernatant as nucleoplasm fraction.

10. Suspend the digested pellet in 500 μL of Buffer A.

11. Purify the suspension using sucrose density gradients (from top to bottom: 0.25–1.6–2.4 M) (*see* **Note 26**).

12. Centrifuge for 20 min at $100,000 \times g$.

13. Collect sucrose fractions (~200–500 μL/fraction on average).

14. Perform immunoblotting by loading 25–35 μL/fraction (*see* **Note 27**).

15. Obtain purified INM-sucrose fractions at the 0.25–1.6 M sucrose interface (Fig. 4, *see* **Note 28**).

16. Resuspend another set of digested pellet in NETN buffer containing phosphatase and protease inhibitors instead of Buffer A as described in Subheading 3.2, **step 10**.

17. Sonicate the NETN-resuspension using Bioruptor sonication device (*see* **Note 17**) or Ultrasonic Processor (*see* **Note 18**).

18. Centrifuge the NETN-resuspension for 20 min at $16,100 \times g$ at 4 °C.

19. Collect the supernatant as the INM-enriched portion.

20. Perform immunoblotting to validate the purity of various fractions, including nonnuclear, ONM-enriched, INM-enriched, and NP, by loading 50 μg/well of proteins (Fig. 5, *see* **Note 29**).

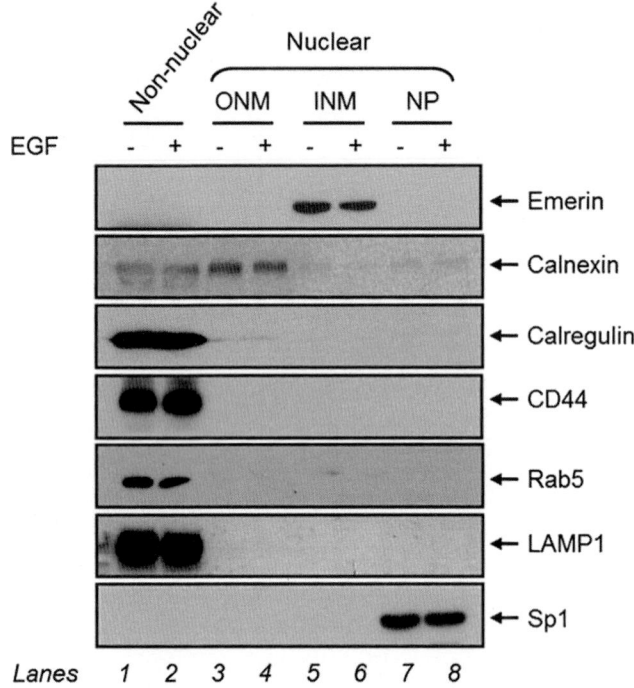

Fig. 5 Detection of cross-contamination of the INM fractions. Biotinylated cell surface proteins were isolated using cellular fractionation and subjected to immunoblotting with the indicated antibodies. This research was originally published in ref. 13 and is reproduced with permission from the American Society for Biochemistry and Molecular Biology

21. Subject the INM-enriched fraction purified from Subheading 3.2, **step 20**, to immunoprecipitation using streptavidin-agarose beads (Fig. 6, *see* **Note 30**).

4   Notes

1. Fresh fractionation buffer containing protease inhibitors should be prepared each time. If protease inhibitor mixture/cocktail is used, please add it to the final concentration as recommended by the individual manufacturer; that is, Sigma recommends a 1:100 inhibitor-to-buffer ratio for mammalian cell lysates.

2. Including 5 % glycerol in NETN buffer will help protein stabilization and in prevention of protein aggregation.

3. The Dounce tissue grinder/homogenizer is designed primarily for cellular work in which the nucleus remains morphologically intact after homogenization. Different working volumes can be selected (e.g., 2 or 7 mL).

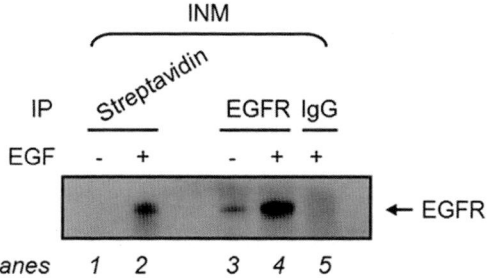

Fig. 6 Cell surface EGFR targets to the INM upon EGF stimulation. The purified INM fractions in Fig. 5 were immunoprecipitated using streptavidin-agarose beads and anti-EGFR antibodies. Immunoprecipitation performed with IgG served as negative control. This research was originally published in ref. 13 and is reproduced with permission from the American Society for Biochemistry and Molecular Biology

4. Similar to the previous note, cell lysis can be performed in different sizes of centrifuge tubes (1.5 and 15 mL).

5. Sodium citrate solution in Buffer A can be prepared prior to use and stored at 4 °C. However, fresh Buffer A/protease inhibitor mixture should be prepared each time.

6. We recommend preparing fresh Sulfo-NHS-LC Biotin solution before use.

7. We find that 0.25 M sucrose is frequently used in the procedures of subnuclear fractionation; therefore, it will be more convenient to make it in a larger volume by dissolving 43 g sucrose in 500 mL water and storing it at 4 °C.

8. After dissolving EGF as recommended by the individual manufacturer, store it in aliquots immediately at $\leq$–20 °C to avoid repeated freeze-thaw cycles.

9. We find that anti-emerin Ab-1 antibody (Thermo Scientific) is also applicable for immunoblotting in addition to immunohistochemistry recommended by the manufacturer.

10. Cell confluency should not reach 100 % before EGF treatment. Cells such as A431 respond best to EGF when at confluency of 60–70 % (about 1×10^7 cells).

11. The volume of lysis buffer used in the assay can be scaled up accordingly, e.g., 300 µL/6 cm culture dish and 800 µL/10 cm culture dish. Cells will generally detach from the dish after adding lysis buffer to cells and gentle shaking for 5–10 s. We recommend placing them on ice or in the 4 °C refrigerator if scraping a large number of plates.

12. This step allows cells packed in lysis buffer containing the detergent (Nonidet P-40) to swell prior to Dounce homogenization.

13. As mentioned in **Note 3**, collecting cell lysate at a larger buffer volume using the 7 ml Dounce homogenizer is recommended to increase work efficiency. For a 7 mL Dounce homogenizer with a tight pestle, a maximum of 2 mL of cells can be homogenized for each cycle.

14. Perform Dounce homogenization gently. If cells are homogenized quickly with up-and-down strokes, a thin white layer after centrifugation may be present above the supernatant. If this happens, transfer the supernatant carefully to exclude the top white layer into a new microcentrifuge tube as a nonnuclear/cytoplasmic fraction. Combine both the bottom pellet and the remaining white layer as a new nuclear pellet for subsequent procedures.

15. Insufficient homogenization may lyse cells incompletely, resulting in contamination of nonnuclear/cytoplasmic proteins in the nuclear fraction; in contrast, excess homogenization may break up nuclei and lead to nuclear protein contamination in the nonnuclear/cytoplasmic fraction. Inappropriate homogenization will lead to inconsistent results. Therefore, a trypan blue exclusion assay to check for cell viability and nuclear integrity after Dounce homogenization is highly recommended, especially when fractionating a cell line for the first time. In brief, take an aliquot of the homogenized cells, mix with equal volume of trypan blue, and then examine the cells under a microscope. Trypan blue should be excluded from intact cells but taken up inside lysed cells, which are considered nonviable. As shown in Fig. 1, after cell lysis and homogenization following the procedures described in this chapter, around 98 % cells were lysed as validated by trypan blue staining; meanwhile, the nuclei of lysed cells were still morphologically intact.

16. The pellet as packed nuclei is easily dispersed by gently pipetting if the majority of nuclei are intact. However, a sticky lump resulting from DNA released from broken nuclei may occur after pipetting. In such a case, an optimal homogenization condition, such as number of Dounce strokes, should be determined for each individual experimental cell line as described in **Note 15**.

17. The volume of NETN buffer used in the assay can be scaled up accordingly. A smaller volume of NETN buffer can be used to increase the final concentration of nuclear extract but may produce foams during the sonication process. We find that Bioruptor sonication device (cycle number: 10; time: 30 s) works well for such smaller volumes ≤500 μL to prevent foaming. Adjust the parameters if required. If a larger sample volume is required, an appropriate Bioruptor material suitable for larger volumes should be replaced or Ultrasonic Processor should be used as mentioned below.

18. A larger volume of fraction (>500 μL) can be sonicated by Ultrasonic Processor with the parameter setting of amplitude at 30 and 10-s sonication with 1-min rest interval; repeat for a total of five cycles. Prior to sonication, we recommend incubating the nuclear pellet in NETN buffer on ice for 5 min to precool the resuspension. Adjust power settings and timing cycles if required. Check that there is 1-min interval between cycles of sonication to prevent sample from overheating, which is indicated by formation of a gray pellet after centrifugation in Subheading 3.1, **step 26**. Check sonication efficiency by observing the nuclear morphology under a microscope using a trypan blue exclusion assay as mentioned previously.

19. The average yield of nuclear fraction from 1×10^7 A431 cells is about 400–500 μg as determined by Bradford protein assay (Bio-Rad, Hercules, CA, USA).

20. The purity of nonnuclear/nuclear fractions can be validated by nonnuclear marker tubulin and nuclear marker lamin B. In the figure shown, we detected an EGF-dependent population of nuclear EGFR and also a significant level of EGFR in the nonnuclear fraction containing proteins extracted from the cell surface membrane where traditional EGFR is mainly located. Loading a 1:5 ratio of nonnuclear to nuclear proteins for immunoblotting is recommended.

21. For MRIN studies, cell surface proteins may be labeled with biotin prior to subcellular fractionation to further demonstrate that nuclear RTK is indeed from the cell surface but not as a result of newly synthesized RTK from the ER. For instance [13], biotinylate cell surface proteins in MDA-MB-468 breast cancer cells with 1 mM Sulfo-NHS-LC-Biotin at room temperature for 30 min and then wash cells three times with the quenching buffer to stop reaction. Treat the biotinylated cells with or without EGF (50 ng/ml) at 37 °C for 30 min, followed by the subcellular and subnuclear fractionation as described in this chapter.

22. Fresh Buffer A/protease inhibitor mixture should be prepared each time. The final concentration can be adjusted accordingly.

23. It has been reported that treatment of isolated nuclei with 1 % (w/v) sodium citrate preferentially removes the ONM of the NE without damaging the INM [17].

24. The ONM-enriched portion can be further purified by centrifugation through sucrose cushion. In brief, subject the resulting supernatant to a 0.25–1.6 M sucrose gradient, and centrifuge for 20 min at $100,000 \times g$. Collect the ONM-sucrose fraction at the 0.25–1.6 M sucrose interface.

25. We find that sonication before adding DNase I efficiently digests nuclei pellet and prevents the formation of viscous DNA.

26. We make a sucrose gradient solution composed of 4 mL each of 2.4, 1.6, and 0.25 M sucrose from bottom to top, which can be scaled up accordingly. A sample volume of 500 μL per tube is added to a gradient volume of approximately 12.5 mL.

27. If the number of collected sucrose fractions exceeds the number of loading wells for one set of immunoblotting, partial sample collection can be loaded instead.

28. The recovery of INM at the sucrose interface represents two major fractions (6 and 7) that can be validated by INM marker emerin. As shown in Fig. 4, EGFR was consistently distributed in the fractions detected by emerin, indicating that INM fractions contain EGFR. In addition, fractions recovered at the 1.6–2.4 M sucrose interface may include nucleolar and undigested nuclear material [17].

29. The purity of nonnuclear, ONM-enriched containing the ER membrane, INM-enriched, and NP fractions defined by the ER membrane and lumen markers (calnexin and calregulin), cell surface protein (CD44), early endosome protein (Rab5), late endosome protein (LAMP1), nuclear protein (Sp1), and INM protein (emerin) can be used to validate cross-contamination during subcellular and subnuclear fractionation.

30. Other cellular fractionation methods have also been reported. For example, we have used the commercially available ProteoExtract Subcellular Proteome Extraction Kit (EMD Millipore, Billerica, MA, USA) with a slight modification to purify the cytosolic, organelle membrane, and nucleic fractions [18] (Fig. 7). In addition, non-membranous nuclear proteins can also be extracted using high-salt buffer supplemented with 500 mM NaCl and 25 % glycerol from the isolated nuclei; meanwhile, the remaining pellet solubilized in sodium dodecyl sulfate is composed of the nuclear and ER membranes [19]. These methods are convenient for traditional subcellular fractionation but have limitations if the goal is to study protein expression or distribution within the nuclei including the nuclear membranes and nuclear extracts/nucleoplasm. We provide a comprehensive subnuclear fractionation procedure for separating the intact nuclei into the ER membrane, ONM, INM, and nucleoplasm.

Acknowledgements

This study was funded in part by the following grants: National Institutes of Health (CA109311, CA099031, and CCSG CA16672); National Breast Cancer Foundation; the University of Texas MD Anderson-China Medical University and Hospital Sister Institution Fund; Cancer Research Center of Excellence

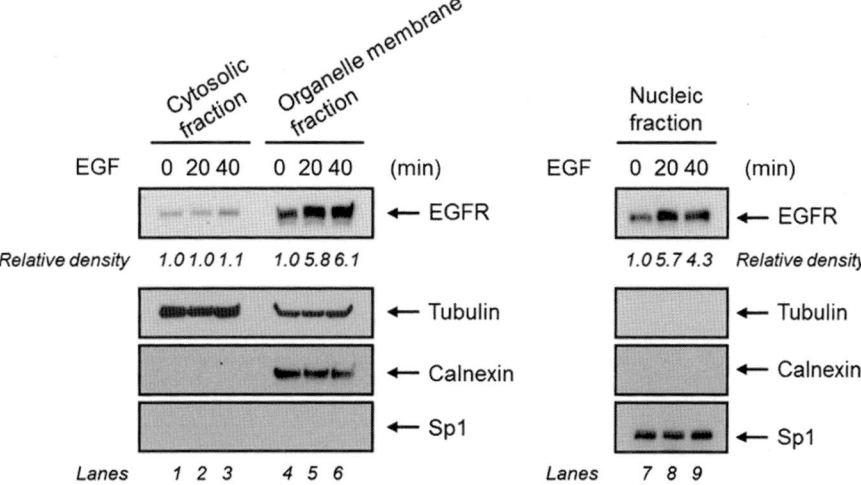

Fig. 7 EGF induces organelle membrane-bound and nuclear EGFR. MDA-MB-468 cells maintained in a serum-starved medium for 24 h were treated with EGF at the indicated time points. Cells were biochemically separated by using the ProteoExtract Subcellular Proteome Extraction kit with a slight modification into various fractions, including cytosolic (*lanes 1–3*), organelle membrane (*lanes 4–6*), and nucleic fractions (*lanes 7–9*), and subjected to immunoblotting with the indicated antibodies. This research was originally published in ref. 18 and is reproduced with permission from the American Society for Biochemistry and Molecular Biology

(MOHW103-TD-B-111-03, Taiwan); Program for Stem Cell and Regenerative Medicine Frontier Research (NSC102-2321-B-039-001, Taiwan); International Research-Intensive Centers of Excellence in Taiwan (NSC103-2911-I-002-303, Taiwan); and the Center for Biological Pathways.

References

1. Davis JR, Kakar M, Lim CS (2007) Controlling protein compartmentalization to overcome disease. Pharm Res 24:17–27

2. Lemmon MA, Schlessinger J (2010) Cell signaling by receptor tyrosine kinases. Cell 141:1117–1134

3. Huo L, Hsu JL, Hung MC (2014) Receptor tyrosine kinases in the nucleus: nuclear functions and therapeutic implications in cancers. In: Kumar R (ed.) Nuclear signaling pathways and targeting transcription in cancer. Springer, New York, NY

4. Wang YN, Hung MC (2013) Nuclear localization of receptor tyrosine kinases in cancers. Canc Clin Res 1:1–2

5. Wang YN, Hsu JL, Hung MC (2013) Nuclear functions and trafficking of receptor tyrosine kinases. In: Yarden Y, Tarcic G (eds) Vesicle trafficking in cancer. Springer, New York, NY, p 159–176

6. Du Y, Hsu JL, Wang YN, Hung MC (2014) Nuclear functions of receptor tyrosine kinases. In: Wheeler DL, Yarden Y (eds) Receptor tyrosine kinase: structure, functions and role in human disease. Springer, New York, NY

7. Wang SC, Hung MC (2009) Nuclear translocation of the epidermal growth factor receptor family membrane tyrosine kinase receptors. Clin Cancer Res 15:6484–6489

8. Fan QW, Cheng CK, Gustafson WC et al (2013) EGFR phosphorylates tumor-derived EGFRvIII driving STAT3/5 and progression in glioblastoma. Cancer Cell 24:438–449

9. Du Y, Shen J, Hsu JL et al (2014) Syntaxin 6-mediated Golgi translocation plays an important role in nuclear functions of EGFR through microtubule-dependent trafficking. Oncogene 33:756–770

10. Wang YN, Hung MC (2012) Nuclear functions and subcellular trafficking mechanisms of

the epidermal growth factor receptor family. Cell Biosci 2:13

11. Carpenter G, Liao HJ (2013) Receptor tyrosine kinases in the nucleus. Cold Spring Harb Perspect Biol 5:a008979

12. Brand TM, Iida M, Li C, Wheeler DL (2011) The nuclear epidermal growth factor receptor signaling network and its role in cancer. Discov Med 12:419–432

13. Wang YN, Yamaguchi H, Huo L et al (2010) The translocon Sec61beta localized in the inner nuclear membrane transports membrane-embedded EGF receptor to the nucleus. J Biol Chem 285:38720–38729

14. Wang ZH, Tian XX, Cheng Y et al (1998) Association of EGFR gene fragments with nuclear matrices in glioblastoma cell lines. Anticancer Res 18:4329–4332

15. Klein C, Gensburger C, Freyermuth S et al (2004) A 120 kDa nuclear phospholipase Cgamma1 protein fragment is stimulated in vivo by EGF signal phosphorylating

nuclear membrane EGFR. Biochemistry 43: 15873–15883

16. Wang YN, Yamaguchi H, Hsu JM, Hung MC (2010) Nuclear trafficking of the epidermal growth factor receptor family membrane proteins. Oncogene 29:3997–4006

17. Humbert JP, Matter N, Artault JC et al (1996) Inositol 1,4,5-trisphosphate receptor is located to the inner nuclear membrane vindicating regulation of nuclear calcium signaling by inositol 1,4,5-trisphosphate. Discrete distribution of inositol phosphate receptors to inner and outer nuclear membranes. J Biol Chem 271:478–485

18. Wang YN, Lee HH, Lee HJ et al (2012) Membrane-bound trafficking regulates nuclear transport of integral epidermal growth factor receptor (EGFR) and ErbB-2. J Biol Chem 287:16869–16879

19. Liao HJ, Carpenter G (2007) Role of the Sec61 translocon in EGF receptor trafficking to the nucleus and gene expression. Mol Biol Cell 18:1064–1072

Chapter 10

Functional G Protein-Coupled Receptors on Nuclei from Brain and Primary Cultured Neurons

Yuh-Jiin I. Jong and Karen L. O'Malley

Abstract

A growing number of G protein-coupled receptors (GPCRs) have been identified on nuclear membranes. In many cases, it is unknown how the intracellular GPCR is activated, how it is trafficked to nuclear membranes, and what long-term signaling consequences follow nuclear receptor activation. Here we describe how to isolate nuclei that are free from plasma membrane and cytoplasmic contamination yet still exhibit physiological properties following receptor activation.

Key words GPCR, Metabotropic glutamate receptor, mGluR5, Nuclear preparation, Calcium imaging, Immunoblotting

1 Introduction

Signal transduction from G protein-coupled receptors (GPCRs) has traditionally been thought to emanate from the cell surface where many signaling complexes are clustered and where extracellular stimuli can interact with GPCR ligand-binding domains. Recently, however, numerous GPCRs have also been found to be associated with various intracellular membranes where, in certain cases, they activate intracellular signaling machinery leading to unique functional responses [1–5]. One such organelle is the nucleus.

The nucleus is typically the largest organelle within the cell being separated from the cytosol by two membranes derived from the endoplasmic reticulum (ER), the inner nuclear membrane (INM), and the outer nuclear membrane (ONM) [6]. Although these membranes are continuous, they are perforated by nuclear pore complexes and by unique protein complexes on either side. For example, proteins on the ONM are similar to those on the adjoining ER whereas the INM is enriched for many specific proteins involved in linking the nuclear envelope to the nuclear lamina and the chromatin [7, 8]. Emerging data underscore the physical connections linking the cytoskeleton, the nuclear membrane, and

Bruce G. Allen and Terence E. Hébert (eds.), *Nuclear G-Protein Coupled Receptors*, Methods in Molecular Biology, vol. 1234, DOI 10.1007/978-1-4939-1755-6_10, © Springer Science+Business Media New York 2015

the nucleoskeleton in order to transmit mechanical forces critical for nuclear movement, cell polarization, and cell motility [7, 8]. Given these critical roles, nuclear envelope proteins have been implicated in a broad range of diseases [9].

A growing number of GPCRs have been demonstrated to be present on both the INM and ONM by ultrastructural, immuno-histochemical, pharmacological, and molecular techniques [1–5]. One such receptor is the metabotropic glutamate receptor, mGluR5, which in previous publications we have shown to function as a nuclear receptor [4, 5, 10, 11]. Specifically, we have found (1) that mGluR5 is highly expressed on inner and outer nuclear membranes of heterologous and physiological cell types [4, 5]; (2) that isolated nuclei expressing mGluR5 respond to permeable or transported agonists by increasing nucleoplasmic Ca^{2+} [4, 5]; (3) that the ligand-binding domain is within the nuclear lumen [4]; (4) that sodium-dependent and -independent transporters/exchangers are present and responsible for transporting mGluR5 agonists across cell surface and nuclear membranes [5, 10, 11]; (5) that endogenous nuclear mGluR5 couples to G_q and PLC to generate IP_3-mediated Ca^{2+} release within the nucleus [10]; and (6) that activation of nuclear mGluR5 generates distinct Ca^{2+} responses as well as downstream signaling cascades separate from cell surface counterparts [11, 12]. These observations and those by others challenge the notion that cells only interact with their environment at the plasma membrane to bring about long-term changes.

Given the growing number of receptors being localized to the nuclear membrane and even within the nucleoplasm itself, this chapter focuses on how to prepare nuclear membranes from brain tissues that are pure yet still capable of carrying out complex signaling processes.

2　Materials

Prepare all solutions using purified deionized water and analytical grade reagents. Prepare and store all buffers at 4 °C. Perform all purification procedures at 4 °C. Make a 50× stock solution of protease inhibitors by dissolving one *complete* protease inhibitor cocktail tablet (Roche Applied Science, Indianapolis, IN) in 1 mL of water and freeze aliquots at −20 °C. Add protease inhibitors to the buffers just prior to use.

- PBS, pH 7.4 (phosphate-buffered saline, Life Technologies, Grand Island, NY).

- TBS (Tris-buffered saline Tween-20: 50 mM Tris, and 150 mM NaCl at pH 7.5).

- TBST (Tris-buffered saline Tween-20: 50 mM Tris, 150 mM NaCl, and 0.1 % Tween-20 at pH 7.5).

Table 1
Preparation of buffers from stock solutions

Final concentration	Stock solution	M.W.	Buffer A (500 mL)	Buffer N (500 mL)	2 M sucrose buffer (250 mL)	1.1 M sucrose buffer (250 mL)
10 mM HEPES, pH 7.5	2.5 M		2 mL	2 mL	1 mL	1 mL
2 mM MgCl₂	1 M		1 mL	1 mL	0.5 mL	0.5 mL
25 mM KCl	1 M		12.5 mL	12.5 mL	6.25 mL	6.25 mL
Buffer N 250 mM sucrose		342.3		42.8 g		
2 M sucrose buffer		342.3			171.2 g	
1.1 M sucrose buffer						94.1 g

- Intracellular buffer (125 mM KCl, 2 mM KH_2PO_4, 2 mM $MgCl_2$, 0.3 mM $CaCl_2$, 10 mM D-glucose, and 40 mM HEPES, pH 7.0).
- Blocking buffer (1 % BSA, 0.25 % milk powder, 0.3 % Triton X-100 in TBS pH 7.4).
- Hypotonic buffer A: 2.0 mM $MgCl_2$, 25 mM KCl, 10 mM HEPES (pH 7.5), and protease inhibitors (*see* Table 1).
- Isotonic buffer N: 2.0 mM $MgCl_2$, 25 mM KCl, 10 mM HEPES (pH 7.5), 250 mM sucrose and protease inhibitors (*see* Table 1).
- 2 M Sucrose buffer: 2.0 mM $MgCl_2$, 25 mM KCl, 10 mM HEPES (pH 7.5), 2 M sucrose and protease inhibitors (*see* Table 1).
- 1.1 M Sucrose buffer: 2.0 mM $MgCl_2$, 25 mM KCl, 10 mM HEPES (pH 7.5), 1.1 M sucrose and protease inhibitors (*see* Table 1).
- Single-sided razor blades.
- Dounce homogenizer.

3 Methods

3.1 Preparation of Nuclei from Cultured Cells or Brain Tissue Regions

1. Isolate postnatal or adult rodent brain tissue from selected regions and keep on ice. Alternatively, harvest cultured cells by scraping cells after washing three times with PBS (*see* **Note 1**).

2. Weigh brain tissue or centrifuge down the cultured cell pellets at $1,000 \times g$ for 10 min.

3. Finely mince the brain tissue on ice with a single-sided razor blade.

4. Resuspend the brain tissue or the cell pellets in 12–20 volumes (per gram of tissue or volume of cell pellet) of Buffer A (*see* **Note 2**).

5. Swell the cells for 5–10 min on ice (*see* **Note 3**).

6. Homogenize the cells on ice in a Wheaton glass Dounce homogenizer using 5–15 strokes (*see* **Note 4**) with either a tight-fitting or loose-fitting pestle depending upon the cell type (*see* **Note 5**). Greater than 90 % of the cells should be disrupted, as judged by trypan blue staining.

7. Readjust the osmolarity by adding 2 M sucrose buffer to the homogenate (*see* **Note 6**) to make the final concentration of sucrose in the buffer to be 250 mM (142.9 µL of 2 M sucrose buffer for 1 mL homogenate).

8. Filter the homogenate through two layers of sterile gauze at the bottom of a 5 mL syringe and centrifuge at $1,000 \times g$ for 10 min.

9. Resuspend the nuclear pellet in Buffer N and gently layer the suspension over 2 mL of medium containing 1.1 M sucrose in Buffer A. Centrifuge at $1,000 \times g$ for 10 min. The pellet from this step contains purified nuclei that are ready to be labeled in various ways.

10. Centrifuge the supernatant from **step 8** at $100,000 \times g$ for 30 min to collect the membrane fraction (the pellet: lysosomes, mitochondria, peroxisomes, ER, Golgi, plasma membrane, endosomes) and cytosol (the supernatant).

11. Snap-freeze the samples for binding assays or western blot analysis in liquid nitrogen and store at −80 °C.

12. Follow procedures in Subheading 3.4 for functional assays with isolated nuclei including fluorescent measurements of Ca^{2+} changes.

3.2 Check the Efficiency and Purity of Isolated Nuclei from Cultured Cells

1. Purity of nuclei can be assessed by loading monolayer cells (HEK 293 or cultured neurons) with the cytoplasmic mitochondrial selective stain MitoTracker Red CM-H$_2$XRos (Life Technologies) as well as the DNA-specific vital dye, Hoechst 33342 (Sigma –Aldrich, St. Louis, MO).

2. Dilute the 1 mM MitoTracker Red CM-H$_2$XRos to 500 nM (for primary neuronal cultures) or 250 nM (for transformed cells) in preferred growth medium without serum for each cell type at 37 °C.

3. Incubate the cells for 45 min at 37 °C.

4. Gently wash the cells once with serum-free medium and replace with Hoechst 33342 (0.5 µg/mL in serum-free medium) for staining the nuclei.

5. Incubate for 20 min at 37 °C.

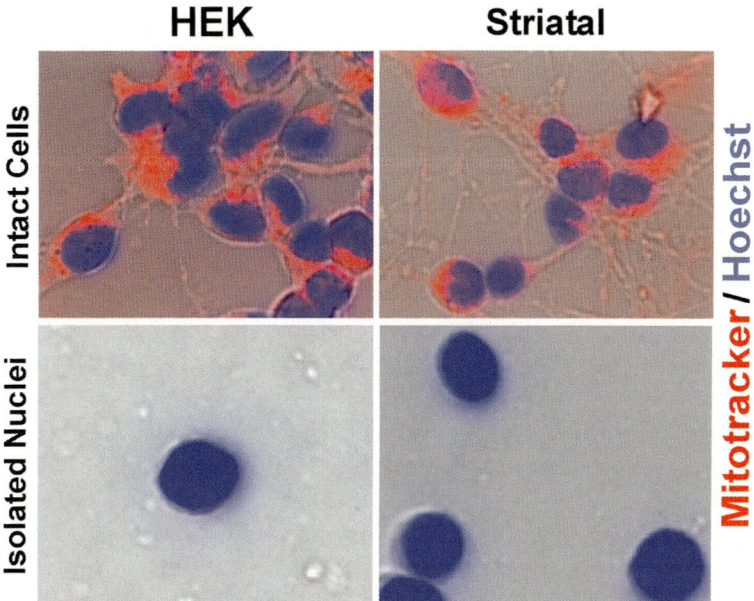

Fig. 1 Using the described nuclear preparation protocol, no MitoTracker Red (excluded from the nucleus) was seen in purified nuclear preparations stained with Hoechst (retained in the nucleus) in either HEK 293 cells or striatal neurons verifying the purity of these organelles

6. Replace with serum-free medium and monitor the mitochondrial and nuclear staining before and after purification procedures (Fig. 1).

3.3 Check the Purity of Isolated Nuclear Brain Tissue via Western Blotting

1. Determine protein concentration of the fractions using the Bradford assay (Bio-Rad Laboratories, Inc., Richmond, CA).

2. Heat aliquots of fractionated proteins at 55 °C for 10 min in sodium dodecyl sulfate polyacrylamide (SDS) sample buffer (*see* **Note 7**).

3. Once lysed, the nuclear pellet can be sticky due to released DNA. Spin the nuclear lysate at $10,000 \times g$ for 5 min to pellet the DNA. Load the appropriate amount of supernatant on an 8.5 % polyacrylamide gel.

4. Separate fractionated proteins by SDS-PAGE and blot onto a Sequi-Blot™ PVDF membrane (Bio-Rad Laboratories, Inc. Hercules, CA).

5. Probe the PVDF membrane with polyclonal anti-mGluR5 (1:1,500; Upstate, Lake Placid, NY), a nuclear membrane marker such as monoclonal anti-lamin B_2 (1:1,000, Life Technologies), and a plasma membrane marker such as polyclonal anti-Pan-Cadherin (1:2,000, Cell Signaling Technology, Inc.) overnight at 4 °C (*see* **Note 8**).

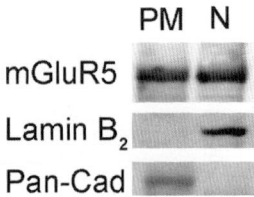

Fig. 2 Using the described nuclear preparation protocol, subcellular fractionation of adult mouse striata shows that mGluR5 can be detected in fractions containing both the nuclear (*N*) and plasma membranes (*PM*). The purity of each fraction is assessed by blotting the same membrane using either the plasma membrane marker pan-cadherin (Pan-Cad) or the nuclear membrane marker lamin B_2

6. Wash the membrane with TBST three times for 15 min each.

7. Probe the membrane with a horseradish peroxidase conjugated with goat anti-rabbit IgG (1:2,000, Cell Signaling Technology, Inc. Danvers, MA) or anti-mouse IgG (1:2,000, Sigma-Aldrich) at room temperature for 1 h.

8. Wash the membrane with TBST three times for 15 min each.

9. Detect the protein bands with enhanced chemiluminescence detection kit (Amersham/GE Healthcare, Piscataway, NJ).

10. Perform densitometric analyses of detected proteins using Molecular Dynamic's Storm 860 Imager (Amersham/GE Healthcare) or ChemiDot™ MP System (Bio-Rad Laboratories, Inc.) (Fig. 2).

3.4 Fluorescent Measurement of Nuclear Ca^{2+} in Isolated Nuclei

1. Load purified nuclei with 20 µM fluorescent Ca^{2+} indicator Oregon Green 488 BAPTA-1AM (Molecular Probes, Eugene, OR), 0.001 % pluronic acid prepared in intracellular buffer together with 1 mM $CaCl_2$, 1 mM ATP, and protease inhibitors for 20–30 min on ice.

2. Wash the nuclei three times with the same buffer minus the Ca^{2+} indicator.

3. Plate the nuclei on poly-D-lysine-coated 35 mm dishes with glass grids.

4. Image immediately using a laser confocal microscope. For example, we use an Olympus BX 50WI together with Olympus LUMPlanFl/IR 40×/0.80w or 60×/0.90w objectives.

5. Capture baseline, real-time images using the Olympus Fluoview FVX confocal laser scanning system together with Fluoview acquisition software (*see* **Note 9**).

6. Add drugs at 100× concentrations to the dish with a gel-loading pipet tip (*see* **Note 10**).

7. Process the images with image analysis software. We use MetaMorph Professional Image Analysis software produced by Universal Imaging (Bedford Hills, New York).

8. The purity of the nuclear preparation and the putative presence of the predicted GPCR can be examined by post hoc identifying nuclei via immunocytochemistry followed by field relocation using the following procedures.

9. Wash the nuclei three times with PBS.

10. Fix the nuclei with 4 % paraformaldehyde (PFA) for 15 min at room temperature.

11. Wash the nuclei three times with PBS (*see* **Note 11**).

12. Block nuclei in blocking buffer for 1 h at room temperature.

13. Incubate the nuclei with polyclonal anti-mGluR5 (1:200, Upstate) overnight at 4 °C (*see* **Note 12**).

14. Wash the nuclei with PBS three times for 10 min each.

15. Incubate the nuclei with goat anti-rabbit Cy3 (1:300; Jackson ImmunoResearch, West Grove, PA) for 1 h at room temperature.

16. Wash the nuclei with PBS three times for 10 min each.

17. Repeat **steps 10–13** with monoclonal anti-Lamin B_2 (1:100, Life Technologies) and goat anti-mouse Alexa 488 (1:300, Life Technologies).

18. Perform post hoc field relocation by an Olympus Fluoview FVX confocal laser scanning system using Fluoview acquisition software (Fig. 3).

4   Notes

1. The single most critical step in purifying healthy nuclei is to rapidly yet gently lyse cells such that nuclei are still intact and not overly exposed to low osmolarity. Therefore, the ratio of tissue weight to Buffer A volume, the length of time cells are exposed to hypotonic solutions, and the number of strokes it takes to homogenize are all important variables. Too little lysis results in whole-cell contamination of the nuclear preparation whereas too much lysis leads to damaged, broken, and unhealthy nuclei.

2. The optimal weight of tissue or volume of cell pellet to volume of Buffer A is 1:10 for cultured HEK 293 or striatal cells, 1:12 for striatum, 1:20 for cortex, and 1:15 for hippocampus.

3. The optimal time period to swell the cells is 10 min for cultured HEK 293 or striatal cells, 5 min for striatum and hippocampus, and 10 min for cortex.

4. The optimal number of strokes is 15 for cultured HEK 293 or striatal cells, 5 strokes for striatum, 10 strokes for cortex, and 7 strokes for hippocampus.

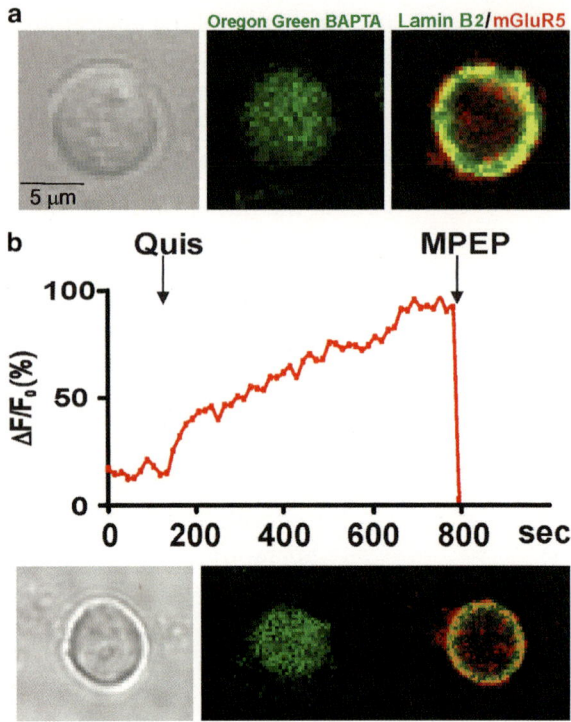

Fig. 3 mGluR5 agonist-mediated Ca^{2+} changes in isolated striatal nuclei. (**a**) *Left panel*: Transmitted light image of purified nucleus from striatal culture loaded with Oregon Green BAPTA (*middle panel*). *Right panel*: mGluR5 (*red*) and lamin B_2 (*green*) co-staining of selected nucleus following drug treatment and post hoc field relocation. (**b**) mGluR5 agonist quisqualate (Quis)-mediated representative trace. Quisqualate application increases nuclear calcium. The mGluR5 antagonist 2-methyl-6-(phenylethynyl) pyridine (MPEP) blocked this response

5. A tight-fitting pestle is best for HEK 293 cells; use a loose-fitting pestle for cultured striatal neurons and brain tissue regions. When purchased, Dounce homogenizers include both tight- and loose-fitting pestles.

6. Nuclei will continue swelling in hypotonic solutions. Therefore, rapidly readjusting the osmolarity once the desired purity is achieved is critical for healthy nuclei. The more swollen the nuclei appear, the more likely it is that they too have been hypotonically shocked. This makes functional recovery unlikely.

7. GPCRs tend to aggregate at high-temperature (100 °C) treatment. To avoid this, samples subjected to SDS-PAGE for GPCR detection are usually treated at 55 °C for 10 min.

8. Alternatively, another plasma membrane marker can be used: monoclonal anti-Na⁺/K⁺-ATPase (α6F, 1:1,000, Developmental Studies Hybridoma Bank, University of Iowa, Iowa City, Iowa).

9. Live images can be captured at 8.59 s/scan for striatal nuclei or 5.36 s/scan for HEK 293 nuclei for 10–15 min.

10. Ca^{2+} changes can be monitored at room temperatures. A physiological chamber with a heating block can be used for temperature control.

11. Fluorescent signals from the Ca^{2+} indicator, Oregon Green 488 BAPTA-1AM, are completely lost following fixation.

12. Alternatively, incubate the nuclei with primary antibody for 2 h at room temperature.

Acknowledgements

This work was supported by National Institutes of Health Grants MH57817 and MH69646 and NS081454. This work was also supported by a FRAXA Research Grant, by the Simons Foundation, and by the Bakewell Family Foundation.

References

1. Tadevosyan A, Vaniotis G, Allen BG et al (2012) G protein-coupled receptor signalling in the cardiac nuclear membrane: evidence and possible roles in physiological and pathophysiological function. J Physiol 590:1313–1330

2. Vaniotis G, Allen BG, Hébert TE (2011) Nuclear GPCRs in cardiomyocytes: an insider's view of β-adrenergic receptor signaling. Am J Physiol Heart Circ Physiol 301:H1754–H1764

3. Bkaily G, Avedanian L, Al-Khoury J et al (2012) Receptors and ionic transporters in nuclear membranes: new targets for therapeutical pharmacological interventions. Can J Physiol Pharmacol 90:953–965

4. O'Malley KL, Jong YJ, Gonchar Y et al (2003) Activation of metabotropic glutamate receptor mGlu5 on nuclear membranes mediates intranuclear Ca^{2+} changes in heterologous cell types and neurons. J Biol Chem 278:28210–28219

5. Jong YJ, Kumar V, Kingston AE et al (2005) Functional metabotropic glutamate receptors on nuclei from brain and primary cultured striatal neurons. Role of transporters in delivering ligand. J Biol Chem 280:30469–30480

6. Isermann P, Lammerding J (2013) Nuclear mechanics and mechanotransduction in health and disease. Curr Biol 23:R1113–R1121

7. Rothballer A, Kutay U (2013) Poring over pores: nuclear pore complex insertion into the nuclear envelope. Trends Biochem Sci 38:292–301

8. Adams RL, Wente SR (2013) Uncovering nuclear pore complexity with innovation. Cell 152:1218–1221

9. Schreiber KH, Kennedy BK (2013) When lamins go bad: nuclear structure and disease. Cell 152:1365–1375

10. Kumar V, Jong YJ, O'Malley KL (2008) Activated nuclear metabotropic glutamate receptor mGlu5 couples to nuclear Gq/11 proteins to generate inositol 1,4,5-trisphosphate-mediated nuclear Ca^{2+} release. J Biol Chem 283:14072–14083

11. Jong YJ, Kumar V, O'Malley KL (2009) Intracellular metabotropic glutamate receptor 5 (mGluR5) activates signaling cascades distinct from cell surface counterparts. J Biol Chem 284:35827–35838

12. Kumar V, Fahey PG, Jong YJ et al (2012) Activation of intracellular metabotropic glutamate receptor 5 in striatal neurons leads to up-regulation of genes associated with sustained synaptic transmission including Arc/Arg3.1 protein. J Biol Chem 287:5412–5425

Chapter 11

Automated Microscopy of Cardiac Myocyte Hypertrophy: A Case Study on the Role of Intracellular α-Adrenergic Receptors

Karen A. Ryall and Jeffrey J. Saucerman

Abstract

Traditional approaches for measuring cardiac myocyte hypertrophy have been of low throughput and subjective, limiting the scope of experimental studies designed to understand it. Here, we describe an automated image acquisition and analysis platform for studying the dynamics of cardiac myocyte hypertrophy in vitro. Image acquisition scripts record 5×5 mosaic images of fluorescent protein-labeled neonatal rat ventricular myocytes from each well of a 96-well plate using the microscope's automated stage and focus. Image analysis algorithms automatically segment myocyte boundaries, track myocytes, and quantify changes in shape. We describe each step of the image acquisition and analysis algorithms and provide specific examples of how to implement them using Metamorph and CellProfiler software. With this system, shape dynamics of thousands of individual cardiac myocytes can be tracked for up to a week. This imaging platform was recently applied to study reversal of cardiac myocyte hypertrophy following withdrawal of the α-adrenergic agonist phenylephrine. Hypertrophy readily reversed at low but not high levels of α-adrenergic signaling, leading to identification of an intracellular population of α-adrenergic receptors responsible for this reversibility delay.

Key words Cardiac hypertrophy, Automated imaging, α-Adrenergic signaling, High-content screening

1 Introduction

Current in vitro approaches for measuring cardiac myocyte hypertrophy have been of low throughput and qualitative, limiting the size and scope of experiments aimed at exploring underlying mechanisms. Applying high-content imaging methodology [1, 2] to cardiac hypertrophy studies can address these limitations by enabling rapid and reproducible collection of hundreds of images at multiple time points with automated quantification of changes in cell size and shape. High-content screening shortens the time between finishing an experiment and obtaining the quantitative results, allowing for faster turnaround in experimental planning.

Bruce G. Allen and Terence E. Hébert (eds.), *Nuclear G-Protein Coupled Receptors*, Methods in Molecular Biology, vol. 1234, DOI 10.1007/978-1-4939-1755-6_11, © Springer Science+Business Media New York 2015

Here, we provide a detailed protocol for a live-cell automated imaging approach to quantify changes in size and shape of cultured cardiac myocytes. This approach can track changes in thousands of individual cardiac myocytes over several days. As a case study, below we describe how this approach was recently applied to study of the role of intracellular α-adrenergic receptors in the dynamics of reversal of phenylephrine (PE)-induced myocyte hypertrophy [3]. This method has also been applied to study the regulation of cardiac myocyte hypertrophy by a Ras/Erk/JNK/p38 network [4] and nuclear protein kinase A [5].

2 Materials

1. Neonatal Cardiomyocyte Isolation Kit, Cellutron cat# nc-6031.

2. SureCoat, Cellutron, Cat# sc-9035, a proprietary blend of collagen and laminin.

3. CellBIND 96-well, flat, clear bottom, black polystyrene, microplates, Corning, cat# 3340.

4. Cell culture media: Dulbecco's modified Eagle media, 17 % M199, 10 % horse serum, 5 % fetal bovine serum, 100 U/mL penicillin, and 50 mg/mL streptomycin.

5. Serum-free cell culture media: Dulbecco' modified Eagle media, 19 % M199, 1 % ITSS, 100 U/mL penicillin, and 50 mg/mL streptomycin.

6. Lipofectamine 2000, Invitrogen.

7. GFP driven under a cardiac myocyte-specific troponin T promoter [6], DNA plasmid, Genscript.

8. Olympus IX81 motorized inverted microscope.

9. Proscan motorized microscope stage (Prior Scientific, Rockland, MA).

10. Olympus UPlanSApo 10×/0.40 numerical aperture (NA) objective.

11. 480/40-nm excitation filter and 535/50-nm emission filter, Chroma filters.

12. Orca-AG CCD camera (Hamamatsu Photonics, Bridgewater, NJ).

13. Microscopy automation image acquisition software, such as Metamorph (Molecular Devices LLC, Sunnyvale, CA).

14. Image J software.

15. Quantitative image analysis algorithm, developed with programs such as CellProfiler [7] (open-source software for automated image analysis, no prior programming knowledge required) or Matlab.

16. Phenylephrine, an α-adrenergic receptor agonist, Sigma-Aldrich.

17. Prazosin, an α-adrenergic receptor antagonist, can be transported inside the cell, Sigma-Aldrich.

18. CGP-12177a, a hydrophilic α-adrenergic receptor antagonist that cannot enter the cell and acts only at the sarcolemma [8], Sigma-Aldrich.

3 Methods

3.1 Cell Culture

1. Isolate rat cardiac myocytes from 1- to 2-day-old Sprague Dawley rats using the Cellutron Neonatal Cardiomyocyte Isolation kit.

2. Plate myocytes on 96-well plate coated with SureCoat, 100,000 myocytes per well.

3. Culture myocytes in cell culture media at 37 °C and 5 % CO_2.

4. 2 days after isolation, transfect cardiac myocytes with 0.4 μg GFP under a troponin T promoter per well using Lipofectamine 2000 according to the manufacturer's instructions (*see* **Note 1**). Wait for 48 h after transfection for adequate GFP expression for imaging (*see* **Note 2**).

5. After collecting initial images, rinse the myocytes and then culture myocytes in cell culture media without serum with the desired perturbation (*see* **Note 3**).

6. Record follow-up images at the desired time points. Myocytes can be reliably imaged for about a week.

7. A complete cell culture timeline for transfection and imaging is shown in Fig. 1.

3.2 Image Acquisition

Software algorithms were developed to automatically focus and collect 25 images in each well of interest in a 96-well plate (Fig. 2). These 25 images are put together into a composite 5×5 mosaic image for each well. By programming the automated microscope stage, the stage moves to the center of each well automatically.

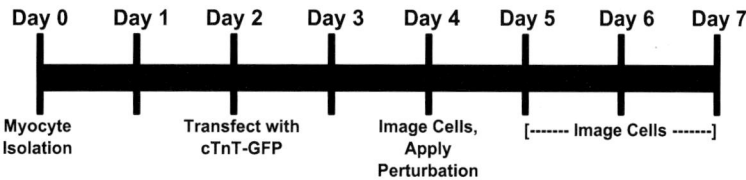

Fig. 1 Timeline for cell culture and imaging for automated imaging of changes in cardiac myocyte shape

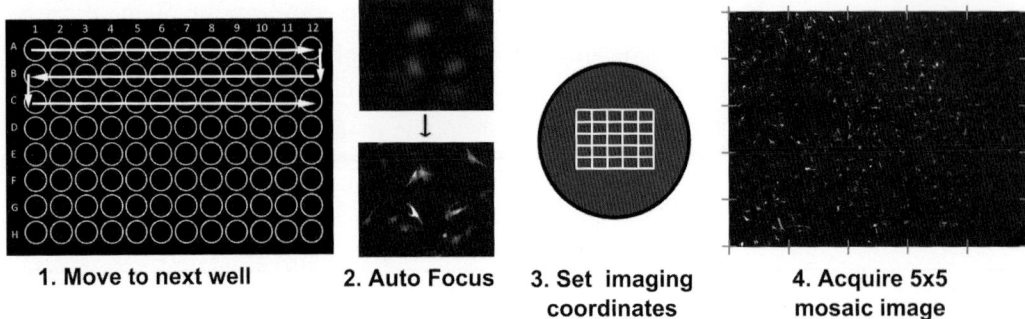

1. Move to next well   **2. Auto Focus**   **3. Set imaging coordinates**   **4. Acquire 5x5 mosaic image**

Fig. 2 Scripts to control the microscope automated stage allow for rapid and reproducible imaging of the same set of myocytes over time. Using pre-saved coordinates for the centroid of each well, the automated stage is programmed to move to each well of interest in a 96-well plate. After (*1*) moving to the next well, the algorithm will (*2*) autofocus, (*3*) set imaging coordinates depending on the size of the mosaic image, and (*4*) acquire the mosaic image. A representative 5×5 mosaic image is shown

Many microscopy image acquisition software packages allow users to develop scripts to automate experiments. The basic workflow for scanning 96-well plates to measure changes in cell shape is as follows:

1. Save coordinates of centroids of each well of interest in your 96-well plate.

   (a) Metamorph has a "scan multi-well plate" command which allows user to specify the dimension of the plate and then select which wells the user wants to image. Save the stage position coordinates of the first well to standardize the path of the scan. The user would then move the stage to this saved location at the beginning of each experiment.

   (b) If your software package does not have a similar built-in function, you can create an array of stage position coordinates, saving the centroid locations of each well.

2. Using these well coordinates, develop a script that moves the stage to each location you want to image. Moving the stage in a serpentine pattern allows for the least amount of total stage movement.

   (a) Again, this can be done in Metamorph using the "scan multi-well plate" function with the plate dimensions.

3. Set an image acquisition script to execute in each well. In Metamorph, these script files are called journals. The script should have the following commands:

   (a) *Coarse autofocus*—Move objective to five different heights, 10 μm apart (±20 μm from the current objective height), and automatically select the height producing the image with the sharpest image resolution. In order for this algorithm to be successful, the focus must be approximated by the user at the beginning of the experiment.

- In Metamorph this operation can be accomplished using the "Adjust Focus" command in the Journal Editor.

(b) *Fine autofocus*—Move objective to five different heights, 2 μm apart (±4 μm from the current objective height), and automatically select the height producing the image with the sharpest image resolution.

- In Metamorph this operation can be accomplished using the "Adjust Focus" command in the Journal Editor with different parameter values.

(c) *Set parameters for the imaging bounds within the well.* If you would like to collect more than one image per well, the user needs to set up the boundary coordinates for imaging within the well.

- In Metamorph this can be done using the "Assign Variable" command. For a 5×5 image area:
    - Set "ScanSlide.ScanUpperLeft.X" to "Stage Position.StageX"-1077.
    - Set "ScanSlide.ScanLowerRight.X" to "Stage Position.StageX" + 1077.
    - Set "ScanSlide.ScanUpperLeft.Y" to "Stage Position.StageY" + 800.
    - Set "ScanSlide.ScanLowerRight.Y" to "Stage Position.Stagex"-800.

These numbers can be modified for smaller or larger imaging areas. Larger imaging areas generate more cell data per condition, but require more computing resources to analyze due to the larger file sizes.

(d) *Collect images* using the set parameters.

- In Metamorph this can be done using the "Scan Slide" command with the above variable set.

- Before first using Scan Slide, this command needs to be calibrated. Also, image acquisition settings are set in this command such as exposure time and binning. An exposure time of 120 ms works well here. These settings can be saved and loaded before beginning image acquisition.

(e) *Save* collected mosaic image.

- In Metamorph this can be done using the "Save Using Sequential File Name" command.

Cell size, shape, and fluorescence intensity data from these mosaic images can then be extracted using automated image analysis pipelines. Since myocyte migration is minimal and the coordinates of the 96-well plate are fixed, we can use these scripts to image the same set of myocytes over time. Therefore, we can measure fold changes in shape of thousands of individual myocytes after perturbations. Myocytes can be reliably imaged over the time period of about a week.

3.3   Image Analysis

Image analysis scripts can be used to load images, correct for noise, segment myocytes, measure shape and fluorescence intensity, and track myocytes over time (Fig. 3). For labs with more computational experience, the following image analysis procedures can be developed in MATLAB. However, for labs with limited programming experience, the open-source program CellProfiler can be used to develop automated image analysis pipelines.

The basic steps of the image analysis pipeline with instructions for implementing this pipeline using CellProfiler are described below. In CellProfiler, image analysis pipelines are built by piecing together built-in image analysis modules. The settings in each module can then be fine-tuned to accommodate the needs of a given image set. A CellProfiler pipeline file and example images for this protocol are freely available at http://bme.virginia.edu/saucerman under Downloads.

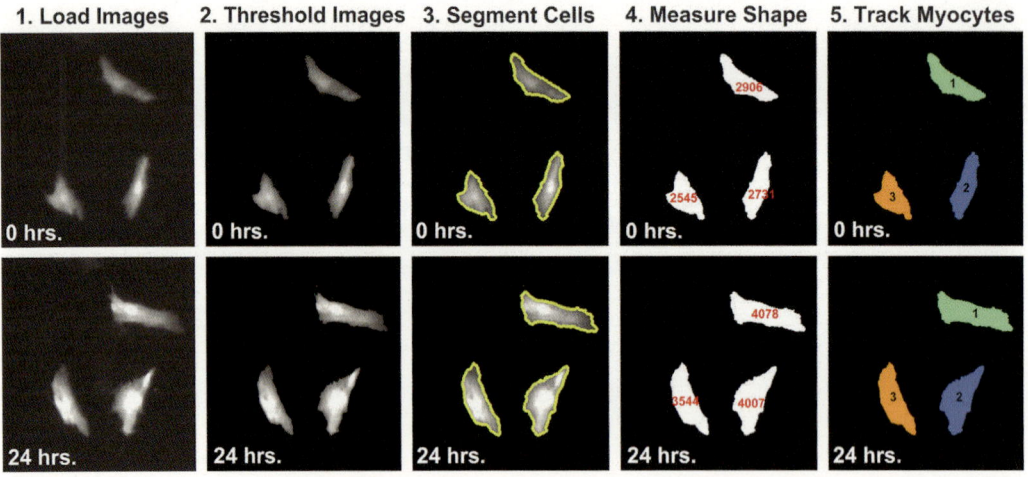

Fig. 3 Image analysis pipelines enable automated quantification of fold changes in shape of cardiac myocytes at multiple time points. The algorithm (*1*) loads images from each time point of a given well, (*2*) thresholds the image to reduce background noise, (*3*) segments myocyte boundaries, (*4*) calculates shape features such as cell area from the segmentation, and (*5*) tracks myocytes over time. Myocytes are tracked using the cell with the closest distance to the centroid at the previous time point

1. *Load images*—Load all images of a given well at each time point measured.

   (a) This can be done using the "Load Images" command in CellProfiler and specifying text labels that these images have in common (such as well number) (*see* **Note 4**).

2. *Background correction*—A pixel intensity threshold for image noise is calculated using the Otsu method [9]. This threshold value is then subtracted from the image to remove background noise.

   (a) Use the "Apply Threshold" command in CellProfiler, and set a noise threshold for the image using the "Otsu Adaptive" thresholding method for two classes. Minimize the weighted variance and use a correction factor of 1 (no correction).

3. *Segment myocytes*—Calculate a pixel intensity that distinguishes cells from background using the Otsu method. To segment any adjacent cells, first calculate the number of myocytes in the image using the number of local maxima in a smoothed image and then use a watershed algorithm [10] to calculate the dividing lines. Since Lipofectamine 2000 transfection efficiency in neonatal ventricular cardiac myocytes is 10–15 %, expressing myocytes are rarely adjacent to each other. Discard myocytes touching the edge of the image from analysis.

   (a) Use the "Identify Primary Objects" command in CellProfiler with the following specifications:

   - Specify the typical diameter range of the myocytes to eliminate dead cells or noise from the analysis. Discard myocytes outside of this range.

   - Use Otsu Adaptive thresholding method, two classes, minimizing weighted variance, with a correction factor of 0.8.

   - Use "Intensity" to distinguish clumps of objects and draw dividing lines. Since myocytes on average are brighter towards the center, GFP fluorescence intensity is useful in determining the number of myocytes in a cluster. The number of local maxima in a smoothed image is used to count the number of myocytes. Then a watershed algorithm is used to set the boundary lines between myocytes.

   - Specify an appropriate maximum number of objects based on your image sizes and transfection efficiency to prevent program crashes on high-noise images.

4. *Overlay outlines on original image*—Create an image overlaying the segmentation result onto the original image. This can be used to check accuracy of myocyte segmentation.

   (a) Use the "Overlay Outlines" command in CellProfiler.

5. *Measure myocyte size and shape*—Use the segmentation result to calculate shape features for each myocyte including area, perimeter, form factor, major and minor axis length, and orientation. For studying hypertrophy, we focused on changes in myocyte area.

   (a) Use the "Measure Object Size Shape" command in CellProfiler.

6. *Measure myocyte GFP fluorescence intensity*—Measure the integrated fluorescence intensity of each myocyte based on the segmentation. Looking at integrated intensity reveals information about the amount of expression of the plasmid. Certain perturbations may increase the expression depending on the promoter.

   (a) Use the "Measure Object Intensity" command in CellProfiler.

7. *Track myocytes between days*—Give each myocyte an identification number and then find each myocyte's position in images at later time points by identifying the closest myocyte to its original position (*see* **Note 5**).

   (a) Use the "Track Objects" command in CellProfiler with "Distance" as the tracking method. This gives each myocyte an identification number and identifies which myocyte in subsequent images is the same cell. The pixel neighborhood the algorithm searches around each myocyte on later days is specified. 150 pixels is a good starting value. These identification numbers then can be used to sort the myocyte measurements in post-processing.

8. *Save image processing images*—Save images from desired points in image processing algorithm such as the image with the segmentation outlines.

   (a) Use the "Save Images" command in CellProfiler to save the image with cell outlines and the image with cell ID numbers from the tracking algorithm.

4   Case Study: Using Automated Imaging to Study the Role of Intracellular α-Adrenergic Receptors on Reversal of Hypertrophy

Since our automated image acquisition and analysis pipeline enabled efficient tracking of myocytes over time, we were able to perform the first (to our knowledge) study of reversal dynamics of cardiac myocyte hypertrophy. After collecting initial myocyte images, we stimulated neonatal rat ventricular myocytes with each of the five increasing concentrations of α-adrenergic agonist phenylephrine (PE) in serum-free culture media. After 24 h with PE, myocytes were imaged, rinsed, and cultured in serum-free cell culture media. Follow-up images were recorded 24 and 48 h after PE washout.

We calculated fold change in myocyte area at each time point with respect to initial myocyte size. Due to significant cell-to-cell variability, tracking changes in features of individual cells substantially decreased dispersion in the data. Approximately 400 myocytes were analyzed per condition, combining data from three separate myocyte isolations. Segmenting and tracking this many myocytes over 4 days would be infeasible without automated methods.

Surprisingly, changes in myocyte area after washout of PE showed a concentration-dependent delay in reversal of hypertrophy. After PE washout, myocytes given lower concentrations of PE (≤ 10 µM) immediately began decreasing in size while myocytes given the highest concentrations of PE (100 µM and 1 mM) continued to increase in size (Fig. 4a).

We hypothesized that the delayed reversal of myocyte hypertrophy was caused by sustained α-adrenergic signaling. To test this hypothesis, we used two different α-adrenergic receptor antagonists after washout: prazosin, an antagonist that can act at the sarcolemma or be transported inside the cell, and CGP-12177a, a membrane-impermeable antagonist that can therefore only act at the sarcolemma [8]. Administration of prazosin after PE washout immediately began to reverse hypertrophy at all concentrations of PE, while myocytes given CGP-12177A continued to have a concentration-dependent delay in reversal of hypertrophy (Fig. 4b). These results indicated that signaling at the level of the α-adrenergic receptor (rather than downstream signaling) was responsible for sustaining hypertrophy with high concentrations of PE.

Due to differences in membrane permeability of prazosin and CGP-12177a and previous reports of intracellular α-adrenergic receptors [11, 12], we hypothesized that the reversal delay was caused by intracellular buildup of PE acting on intracellular α-adrenergic receptors. We developed mathematical models of several potential scenarios, finding that only models incorporating intracellular receptors could predict the observed data. Follow-up experiments with BODIPY-FL fluorescently labeled prazosin confirmed the presence of intracellular α-adrenergic receptors in unstimulated cardiac myocytes. Further, PE or phentolamine displaced fluorescent prazosin from intracellular α-adrenergic receptors [3].

Overall, the high-throughput live-cell image acquisition and analysis approach played a critical role enabling quantitative study of hypertrophy reversal with transient PE. These data together with mathematical modeling and receptor ligand imaging indicated a key role of intracellular α-adrenergic receptors in the initiation and reversal of PE-induced myocyte hypertrophy. As intracellular α-adrenergic receptors have also been reported in the adult mouse heart [11], further studies will be needed to examine whether their localization helps to sustain cardiac hypertrophy in vivo.

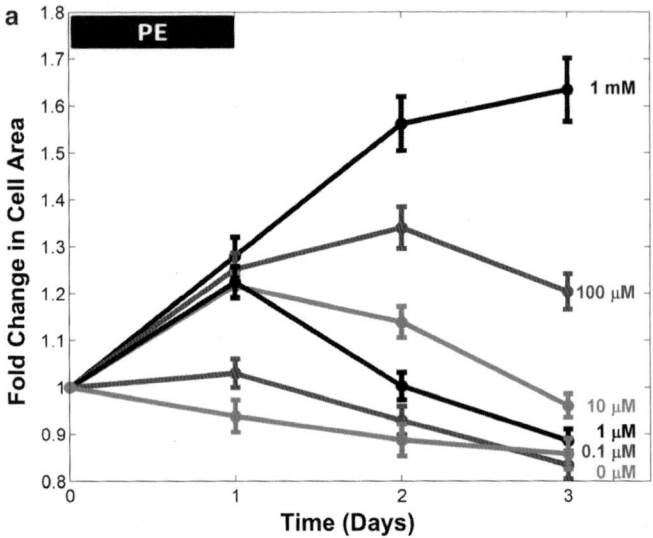

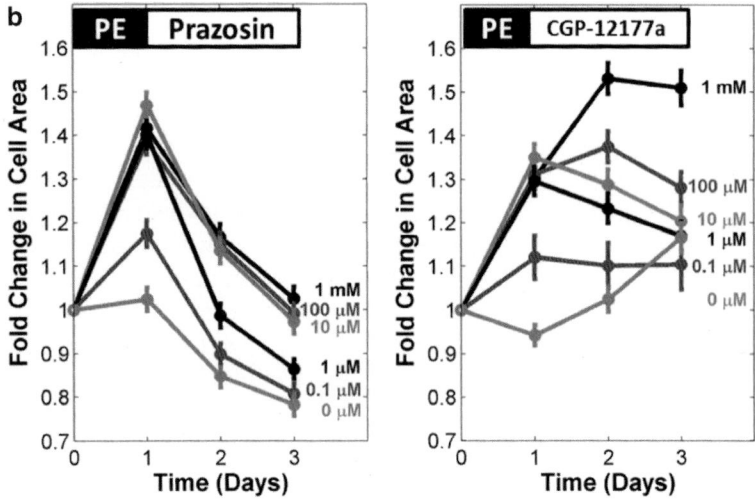

Fig. 4 (a) Fold change in myocyte area during and after 24-h treatment with each of the five doses of α-adrenergic receptor agonist PE. Reversal of cell area after PE washout exhibited a marked concentration-dependent reversibility delay. Myocytes treated with low concentrations of PE readily decreased in area after PE washout, while myocytes treated with high concentrations of PE ($\geq$100 μM) continued to increase in size. Error bars are ±SE. (b) Fold change in myocyte area after 24-h treatment with one of the five doses of PE. After 24-h PE treatment, myocytes were given 10 μM prazosin, a membrane-permeable α-adrenergic receptor antagonist (*left*), or 10 μM CGP-12177a, a membrane-impermeable α-adrenergic receptor antagonist that can only act at the sarcolemma (*right*). Prazosin accelerated reversal of myocyte area after PE washout, while CGP-12117a did not affect reversal of hypertrophy. Error bars are ±SE. These data together with a mathematical model and additional experiments support the hypothesis that internalized PE acts on intracellular α-adrenergic receptors to sustain hypertrophy. Reprinted from ref. 3 with permission from Elsevier

5   Notes

1. Myocytes may also be transfected 24 h after isolation with similar results.

2. While GFP fluorescence intensity increases over the course of the experiment, these increases have minimal effects on area measurements (<5 %) [3].

3. After initial images are collected, it is important to wash and transfer myocytes to media without serum since serum causes high levels of myocyte growth.

4. Depending on the file type the images are saved as by the image acquisition software, the images may need to be converted to a different file type. This can be done using the batch convert plug-in in ImageJ software. CellProfiler can read .png, .tif, .bmp, and .jpg.

5. Myocytes migrate minimally and can be robustly tracked using distance from the previous location. However, if follow-up images are >~100 pixels out of alignment, the tracking algorithm will fail. Therefore, image alignment should be checked and corrected before running the algorithm. If images are dramatically out of alignment, preprocess images in a new CellProfiler pipeline using the "Align" module or align by batch cropping images from each time point in ImageJ.

Acknowledgements

This work is supported by the National Science Foundation (predoctoral fellowship to K.R., CAREER grant to J.S.) and the National Institutes of Health (grant HL094476).

References

1. Zanella F, Lorens JB, Link W (2010) High content screening: seeing is believing. Trends Biotechnol 28:237–245

2. Evans JG, Matsudaira P (2007) Linking microscopy and high content screening in large-scale biomedical research. Methods Mol Biol 356:33–38

3. Ryall KA, Saucerman JJ (2012) Automated imaging reveals concentration dependent delay in reversibility of cardiac myocyte hypertrophy. J Mol Cell Cardiol 53:282–290

4. Ryall KA, Holland DO, Delaney KA et al (2012) Network reconstruction and systems analysis of cardiac myocyte hypertrophy signaling. J Biol Chem 287:42259–42268

5. Yang JH, Polanowska-Grabowska RK, Smith JS et al (2014) PKA catalytic subunit compartmentation regulates contractile and hypertrophic responses to β-adrenergic signaling. J Mol Cell Cardiol 66:83–93

6. Prasad K-MR, Xu Y, Yang Z et al (2011) Robust cardiomyocyte-specific gene expression following systemic injection of AAV: in vivo gene delivery follows a Poisson distribution. Gene Ther 18:43–52

7. Carpenter AE, Jones TR, Lamprecht MR et al (2006) Cell profiler: image analysis software for identifying and quantifying cell phenotypes. Genome Biol 7:R100

8. Staehelin M, Simons P, Jaeggi K et al (1983) CGP 12177: a hydrophilic β-adrenergic receptor

radioligand reveals high affinity binding of ago- nists to intact cells. J Biol Chem 258:3496–3502

9. Otsu N (1979) A threshold selection method from gray-level histograms. IEEE Trans Syst Man Cybern 9:62–66

10. Vincent L, Soille P (1991) Watersheds in digi- tal spaces: an efficient algorithm based on immersion simulations. IEEE Trans Pattern Anal Mach Intell 13:583–597

11. Wright CD, Chen Q, Baye NL et al (2008) Nuclear α1-adrenergic receptors signal acti- vated ERK localization to caveolae in adult car- diac myocytes. Circ Res 103:992–1000

12. Pediani JD, Colston JF, Caldwell D et al (2005) β-arrestin-dependent spontaneous α1a- adrenoceptor endocytosis causes intracellular transportation of α-blockers via recycling com- partments. Mol Pharmacol 67:992–1004

Chapter 12

Measuring Intranuclear and Nuclear Envelope [Ca^{2+}] vs. Cytosolic [Ca^{2+}]

Senka Ljubojević and Donald M. Bers

Abstract

Nuclear Ca^{2+} regulates key cellular processes, including gene expression, apoptosis, assembly of the nuclear envelope, and nucleocytoplasmic transport. Quantification of subcellularly resolved Ca^{2+} signals is, therefore, essential for understanding physiological and pathological processes in various cell types. However, the properties of commonly used Ca^{2+}-fluorescent indicators in intracellular compartments may differ, thus affecting the translation of Ca^{2+}-dependent fluorescence changes into quantitative changes of Ca^{2+} concentration. Here, we describe technical approaches for reliable subcellular quantification of [Ca^{2+}] in the cytoplasm vs. the nucleus and the nuclear envelope by in situ calibration of fluorescein-derived fluorescent indicators Fluo-4 and Fluo-5N.

Key words Ca^{2+} concentration, In situ calibration, Nucleus, Nuclear envelope, Cytoplasm, Fluo-4, Fluo-5N

1 Introduction

The intranuclear calcium concentration ([Ca^{2+}]$_{nuc}$) regulates key cellular processes, including gene expression in various cell types, and its alterations are involved in cellular remodeling processes [1, 2]. Intranuclear Ca^{2+} fluxes follow different kinetics and may be—upon specific stimuli—regulated independently from cytoplasmic Ca^{2+} fluxes [3, 4], due to the insulation of the nucleus from the surrounding cytoplasm by the nuclear envelope (NE). The NE not only controls nuclear structural integrity, but also bidirectional transport of ions (including Ca^{2+}) and macromolecular cargo and acts as a functional Ca^{2+} store to regulate intranuclear Ca^{2+} concentration and Ca^{2+}-dependent signaling and gene expression [5]. The nuclear envelope consists of the inner and outer nuclear membrane, and is interrupted by numerous macromolecular nuclear pore complexes (NPC) connecting the nucleoplasm and the cytoplasm and allowing the passive diffusion of small molecules and ions between these two compartments. On the other hand, the

Bruce G. Allen and Terence E. Hébert (eds.), *Nuclear G-Protein Coupled Receptors*, Methods in Molecular Biology, vol. 1234, DOI 10.1007/978-1-4939-1755-6_12, © Springer Science+Business Media New York 2015

nuclear envelope expresses Ca^{2+}-regulating proteins including the sarcoplasmic/endoplasmic reticulum Ca^{2+}-ATPase (SERCA), Ca^{2+} release channels, Ca^{2+}-buffering proteins, G protein-coupled receptors (GPCRs), including receptors for ET-1 and angiotensin II, and the entire machinery for IP_3 signaling directed to the nucleoplasm [6–8] suggesting active regulation of $[Ca^{2+}]_{nuc}$ that can be relatively independent of cytoplasmic Ca^{2+}. Thus, intranuclear $[Ca^{2+}]$ is the sum of Ca^{2+} that enters the nucleoplasm by diffusion from the cytoplasm through NPCs and Ca^{2+} that is released from the nuclear envelope in a specific and regulated manner in response to different stimuli. However, little is known about the autonomous regulation and alterations of $[Ca^{2+}]_{nuc}$ homeostasis under different physiological and pathophysiological conditions, thus underlining the importance of better understanding the relationship between $[Ca^{2+}]_{nuc}$, NE Ca^{2+} concentration ($[Ca^{2+}]_{NE}$), and cytoplasmic Ca^{2+} concentration ($[Ca^{2+}]_{cyto}$) and how Ca^{2+}-dependent signaling in the nucleus is regulated. This requires accurate measurements of $[Ca^{2+}]$ in both the nuclear (i.e., intranuclear and NE) and cytoplasmic compartments.

The fluorescein-derived fluorescent indicators Fluo-4 and Fluo-5N (high and very low Ca^{2+} affinity, respectively) are widely used to monitor $[Ca^{2+}]$ in the cytoplasm and nucleoplasm and cellular Ca^{2+} stores such as sarco/endoplasmatic reticulum and nuclear envelope of various cell types. In attempting to determine subcellular $[Ca^{2+}]$ using these indicators, however, one encounters several technical difficulties, including distinct behavior of the Ca^{2+} indicators in different cellular compartments. Namely, the fluorescence properties of the dyes are altered differentially by the cytoplasmic and nucleoplasmic environment, as observed in in situ studies with three different mouse cell lines loaded with either Fluo-3 or Fluo-4 [9]. A comparison of fluorescent Ca^{2+} indicator properties in HeLa cells [10] or adult cardiomyocytes [4] further confirmed distinct characteristics of Fluo-3 and Fluo-4 in the cytoplasmic vs. nucleoplasmic compartment. Further problems can arise from the sequestration of Ca^{2+} indicators into intracellular organelles, such as the endoplasmic reticulum or mitochondria [11, 12]. Indicators have also been reported to leak from the cytoplasm to the extracellular medium facilitated by sarcolemmal anion transporters [13, 14]. Due to all these factors, any quantitative analysis of $[Ca^{2+}]_{nuc}$ or $[Ca^{2+}]_{NE}$ vs. $[Ca^{2+}]_{cyto}$ is not reliable without proper independent determination of the indicator properties in the intranuclear, NE, and the cytoplasmic compartment. Thus, addressing quantitative changes of $[Ca^{2+}]_{nuc}$, $[Ca^{2+}]_{NE}$, and $[Ca^{2+}]_{cyto}$ requires transformation of raw fluorescence signals into calibrated $[Ca^{2+}]$, taking into account the effects of the different subcellular environments on the characteristics of the indicator.

The most straightforward solution for converting fluorescence into Ca^{2+} concentrations is to determine in situ calibration curves.

Such curves account for the effects of the cellular environment on the characteristics of an indicator and, if spatial resolution permits, the properties of an indicator can be measured in different cellular compartments [10]. Controlling [Ca^{2+}] in calibration solutions requires special care and buffering, in part because even distilled and deionized water supplies contain significant Ca^{2+} contamination, as do the analytical grade chemicals that are used to make physiological buffers [15]. Here, we describe methods to determine in situ characteristics of a frequently used high-affinity Ca^{2+} indicator, Fluo-4, and a low-affinity Ca^{2+} indicator, Fluo-5N, in isolated adult cardiomyocytes and their use in reporting and quantifying cytoplasmic/nucleoplasmic and SR/nuclear envelope Ca^{2+} signals in combination with confocal microscopy.

2   Materials

2.1   Electrode Calibration Solutions

1. *10^{-4} M standard Ca^{2+} solution*: Pipet 10 mL of the 0.1 M Ca^{2+} standard solution (Thermo Scientific, Beverly, MA, USA) into a 100 mL plastic volumetric flask (*see* **Note 1**). Dilute to the mark with ultrapure deionized water (*see* **Note 2**) and mix well.

2. *Background standard solution #1* (in mM) (*see* **Note 3**): 130 NaCl, 5.4 KCl, pH 7.4 (with NaOH/HCl). To render it Ca^{2+} free, put a dialyzing tube (pre-boiled four times in ultrapure deionized water) filled with Chelex 100 into the solution (Chelex 100, 200–400 mesh sodium, Bio-Rad, CA, USA). Since the sodium Chelex resin is basic, multiple washes with deionized water are required to approach neutral pH (*see* **Note 4**). Then, transfer a thick slurry into a dialysis tube. For 250 mL of solution use 1–2 mL Chelex 100. Place the solution on a shaker or turn it over a couple of times every day for 1 week (at room temperature). The solution is then Ca^{2+} free.

3. *Low-[Ca^{2+}]-level electrode calibration solutions*: Add 100 mL of background standard solution #1 to a 150 mL plastic beaker. Add increments of the 10^{-4} M standard Ca^{2+} solution (*as described below*, Subheading 3.2) to the beaker to reach the desired final [Ca^{2+}].

4. *Background standard solution #2* (in mM): 200 Cs-glutamate, 0.5 MgCl$_2$, 5 Mg-ATP (add Mg-ATP on the day of the experiment), pH 7.4.

5. *Direct electrode calibration solutions*: Prepare 10^{-2}, 10^{-3}, 10^{-4}, and 10^{-5} M standard Ca^{2+} solutions by serial dilution of 0.1 M Ca^{2+} standard solution (Thermo Scientific, Beverly, MA, USA). For 10^{-2} M Ca^{2+} solution, pipet 10 mL of the 0.1 M standard into 100 mL volumetric flask. Dilute to the mark with background standard solution #2 (*see* **Note 5**) and mix well.

The second standard is similarly diluted to prepare a third standard, and so on, until the desired range of standards has been prepared.

2.2 Preparation of Intact Adult Cardiomyocytes

1. *Normal Tyrode (NT) solution for* **mouse** *cells* (in mM): 140 NaCl, 5 KCl, **1 CaCl₂**, 1 MgCl₂, 10 glucose, and 10 HEPES, pH 7.4.

2. *Normal Tyrode (NT) solution for* **rat** *cells* (in mM): 140 NaCl, 5 KCl, **1.5 CaCl₂**, 1 MgCl₂, 10 glucose, and 10 HEPES, pH 7.4.

3. *Normal Tyrode (NT) solution for* **rabbit** *cells* (in mM): 140 NaCl, 5 KCl, **2 CaCl₂**, 1 MgCl₂, 10 glucose, and 10 HEPES, pH 7.4.

4. *Laminin solution*: Pipet 20 μL of laminin (1 mg/mL in Tris-buffered saline, Sigma-Aldrich Chemie GmbH, Steinheim, Germany) in 980 μL of NT.

2.3 Dye Solutions

1. *Fluo-4 dye solution*: Dissolve acetoxymethylester form of Fluo-4 (Fluo-4AM, Molecular Probes, Leiden, The Netherlands) in Pluronic F-127 (20 % (w/v) in dimethyl sulfoxide (DMSO); Molecular Probes, Leiden, The Netherlands) and add to the NT solution to obtain the final dye concentration of 8 μM (*see* **Note 6**).

2. *Fluo-5N dye solution*: Dissolve acetoxymethylester form of Fluo-5N (Fluo-5NAM, Molecular Probes, Leiden, The Netherlands) in Pluronic F-127 (20 % (w/v) in dimethyl sulfoxide (DMSO) (Molecular Probes, Leiden, The Netherlands) and add to the NT solution to obtain the final dye concentration of 10 μM (*see* **Note 6**).

2.4 Calibration Solutions for Fluo-4 Measurements

1. *EGTA solution* (in mM): 130 NaCl, 5.4 KCl, 0.5 MgCl₂, 1 EGTA (*see* **Note 7**), 15 BDM (2,3 butanedione monoxime) (*see* **Note 8**), 25 HEPES, 0.01 A23187 or 0.001 ionomycin (*see* **Note 9**), 1.8 2-deoxy-D-glucose, 0.01 rotenone (*see* **Note 10**), 0.005 CPA (cyclopiazonic acid), pH 7.4. Keep the solution in a plastic container. Before adding MgCl₂, EGTA, BDM, Ca²⁺ ionophore (ionomycin or A23187), and metabolic inhibitors (rotenone and CPA), which are added directly before measurements, repeat the Chelex 100 Ca²⁺ removal procedure.

2. *CaEGTA solution* is the same as the EGTA solution except that it contains 2 mM CaCl₂ in addition (*see* **Note 11**).

2.5 Calibration Solutions for Fluo-5N Measurements

1. *Relaxing solution* (in mM): 0.1 EGTA, 10 HEPES (pH 7.2), 120 K-aspartate, 1 MgCl₂, 5 Mg-ATP, 10 reduced glutathione, 5 phosphocreatine di-Tris, pH 7.4.

2. *Permeabilization solution* is the same as relaxing solution with 50 mg/mL saponin and 10 mM caffeine to allow [Ca²⁺] equilibration across the SR.

3. *Internal solution* (in mM): 200 Cs-glutamate, 10 EGTA, 10 HEPES (pH 7.2), 0.5 $MgCl_2$, 5 Mg-ATP, 10 reduced glutathione, 5 phosphocreatine di-Tris, U/mL creatine phosphokinase, 8 % dextran (MW 40,000) to prevent swelling on permeabilization, 1 μmol/L FCCP, 1 μmol/L ruthenium red, 2 μmol/L oligomycin, and 8 μmol/L cyclosporine to limit mitochondrial Ca^{2+} uptake, pH 7.4. Free [Ca^{2+}] is set at concentration ranging from 10^{-7} to 10^{-2} M by addition of appropriate amount of standard Ca^{2+} solution (*see* below, Subheading 3.5).

2.6 Ca²⁺ Imaging and Calibration Components

1. *Confocal microscope* (Zeiss LSM 510 Meta confocal microscope, Carl Zeiss, Jena, Germany) equipped with an ×40 oil-immersion objective lens (N.A. 1.3) and an argon-ion laser (excitation and emission wavelengths are 488 nm and >515 nm, respectively).

2. *Electrical stimulator* (Myopacer, IonOptix, Milton, MA).

3. *Rapid-switching superfusion system* (VC-8P valve controller, Warner Instruments, Hamden, CT).

4. *Glass bottom cell culture dishes* (WillCo Wells B.V., Amsterdam, The Netherlands).

5. *Ion-selective electrode (ISE) for Ca^{2+} measurements* (Orion 97-20 ionplus; Thermo Electron Co., Beverly, MA, USA).

6. *Magnetic stirrer.*

3 Methods

Conduct all procedures at room temperature (22–24 °C) unless otherwise specified.

3.1 Cell Preparation

1. Coat glass-bottomed culture dishes by pipetting 100 μL of freshly prepared laminin solution to the center of the dish and leave it to air-dry for at least 2 h (*see* **Note 12**).

2. Plate freshly isolated cardiomyocytes in NT solution (*see* **Note 13**) on laminin-coated glass and allow them to attach to the bottom for at least 30 min.

3. Load the cells with the Ca^{2+}-sensitive fluorescent dye by 30-min incubation in Fluo-4-AM dye solution (at room temperature), followed by 30-min de-esterification time. When using Fluo-5N, incubate cardiomyocytes in Fluo-5N dye solution for 2 h, and allow additional 1.5 h for de-esterification and outward leak of cytosolic indicator, all at 37 °C. After dye loading, wash cells 2× with 1 mL NT solution (*see* **Note 14**).

Table 1
Cumulative [Ca^{2+}] curve for calibrating Ca^{2+}-sensitive electrode

Step	Volume added	Concentration
1	50 μL	50 nM
2	50 μL	100 nM
3	100 μL	200 nM
4	200 μL	390 nM
5	200 μL	600 nM
6	400 μL	980 nM
7	2 mL	2.9 μM

The indicated volumes of a 10^{-4} M standard Ca^{2+} solution are added, in a stepwise manner, to 150 mL plastic beaker containing 100 mL of background standard solution #1 to achieve the indicated Ca^{2+} concentrations. After each step, the electrode potential (in millivolts) is allowed to stabilize and then recorded

3.2 Low-[Ca^{2+}]-Level Electrode Calibration

1. Add 100 mL of background standard solution #1 to a 150 mL plastic beaker.

2. Rinse the electrode with ultrapure deionized water, blot it dry, and place it into the beaker. Stir the solution thoroughly.

3. Add increments of the 10^{-4} M standard Ca^{2+} solution to the beaker to reach the desired final [Ca^{2+}] of 5.0×10^{-8}, 1.0×10^{-7}, 2.0×10^{-7}, 3.9×10^{-7}, 6.0×10^{-7}, 9.8×10^{-7}, and 2.9×10^{-6} M using the steps outlined in Table 1.

4. Record the stable millivolt reading after each increment (*see* **Notes 15** and **16**).

5. Plot the concentration (log axis) against the millivolt potential (linear axis). Prepare a new calibration curve with fresh standards each day.

3.3 In Situ Calibration of Fluo-4

1. Calculate the total [Ca^{2+}] required for obtaining the desired free [Ca^{2+}] in the presence of 1 mM EGTA using the MaxChelator program (http://www.stanford.edu/~cpatton/maxc.html) (*see* **Note 17**).

2. Prepare the calibration solutions with desired free [Ca^{2+}] by mixing known quantities of EGTA and CaEGTA solutions (*see* **Note 18**). We used calibration solutions with free [Ca^{2+}] of 0, 50, 250, 750, 1,500, and 3,000 nM and 1 mM, but using fewer steps is possible (four points are minimum for the construction of the calibration curve).

3. To confirm the free [Ca^{2+}] measure 100 mL of each Ca^{2+} calibration solutions for Fluo-4 measurements into a clean plastic beaker. Rinse the electrode with ultrapure deionized water, blot it dry, and place the electrode into the sample.

4. Stir the solution thoroughly. When a stable reading is displayed, record the mV value.

5. Determine the sample concentration corresponding to the measured potential from the low-level calibration curve.

6. Fill the reservoirs of the superfusion system with calibration solutions and remove any air bubble from the tubing. Adjust the speed of superfusion to approximately 3 mL/min.

7. Stimulate the cells in NT solution at 1 Hz for 2 min. Switch off the stimulation to record resting $[Ca^{2+}]$. We define resting $[Ca^{2+}]$ as $[Ca^{2+}]$ 1 min after cessation of stimulation.

8. Let the cells reach the equilibrium in each calibration solution for 8 min. During this period, record 2D fluorescent images every 2 min (Fig. 1a). Minimal Fluo-4 (F_{min}) is measured during exposure to Ca^{2+}-free calibration solution and maximal Fluo-fluorescence (F_{max}) during exposure to a calibration solution containing a saturating free $[Ca^{2+}]$ of 1 mM.

9. Plot the nucleoplasmic and cytoplasmic Fluo-4 fluorescence versus $[Ca^{2+}]$ (Fig. 1b) and fit the concentration-response curves using the Hill equation: $F = \{(F_{max} - F_{min})/(1 + (K_d/[Ca^{2+}])^n)\} + F_{min}$.

10. Use the constructed curve to determine the in situ Ca^{2+} dissociation constant (K_d) and the Hill coefficient (0) (*see* **Note 19**). The typical values for K_d obtained in our experimental setup for mouse and rat ventricular cardiomyocytes were ~1,100 nM in the cytoplasm and ~1,200–1,300 nM in the nucleoplasm [4].

11. To transform Fluo-4 fluorescence into free $[Ca^{2+}]$ use the equation for nonratiometric dyes: $[Ca^{2+}] = K_d (F - F_{min})/(F_{max} - F)$, where F is the fluorescence intensity at any given time (*see* **Note 20**).

3.4 Direct Electrode Calibration

1. Add 100 mL of the least concentrated standard to a clean beaker and stir the solution thoroughly.

2. Rinse the electrode with deionized water, blot it dry, and place it into the beaker with the least concentrated standard. Wait for a stable reading and adjust the meter to display the value of the standard.

3. Work the procedure up to the most concentrated standard.

4. Record the resulting slope value. The slope should be between 25 and 30 mV per log unit of $[Ca^{2+}]$ when the standards are between 20 and 25 °C.

3.5 In Situ Calibration of Fluo-5N

1. Calculate the total $[Ca^{2+}]$ required to obtain the desired free $[Ca^{2+}]$ in the presence of 10 mM EGTA using the MaxChelator program (http://www.stanford.edu/~cpatton/maxc.html) (*see* **Note 17**). For Fluo-5N measurements chelex-treated solutions are not needed, because of the higher range of $[Ca^{2+}]$ to be measured, which is less influenced by contaminating Ca^{2+}.

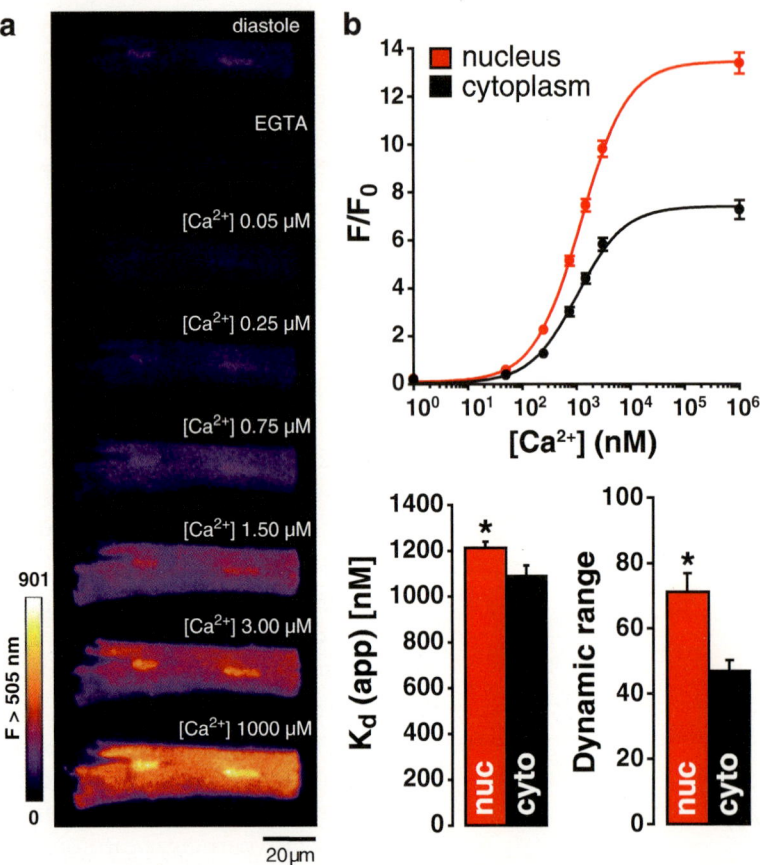

Fig. 1 In situ calibration of fluorescent Ca^{2+} indicator Fluo-4 in the nucleus versus the cytoplasm of mouse ventricular cardiac myocytes (**a**) Original 2D images of Fluo-4 fluorescence of a mouse ventricular myocyte at various $[Ca^{2+}]$ during the calibration protocol. Note that nuclear fluorescence appears brighter than the surrounding cytoplasm, most likely due to the binding of the indicator to nuclear constituents and its accumulation in the nuclear compartment. (**b**) Concentration response curves with seven different $[Ca^{2+}]$ illustrating the in situ Ca^{2+}-dependent fluorescence of Fluo-4 in the nucleus (*red*) versus the cytoplasm (*black*) of mouse ventricular cardiomyocytes (*top*) and apparent dissociation constants for Ca^{2+} binding (K_d (app)) and dynamic range (*bottom*) of Fluo-4 fluorescence in the nucleus (*red*) versus the cytoplasm (*black*). Data in (**b**) from a total of 15 mouse ventricular myocytes. *Asterisks* indicate $P < 0.05$ versus cytoplasm. From Ljubojevic et al. [4]. In situ calibration of nucleoplasmic versus cytoplasmic Ca^{2+} concentration in adult cardiomyocytes. Biophys J 100, 2356–2366. Reprinted from ref. 4 with permission from Elsevier

2. Prepare the calibration solutions with desired free $[Ca^{2+}]$ (e.g., 10^{-2}, 10^{-3}, 10^{-4}, 10^{-5}, and 10^{-7} M) by adding calculated amount of 0.1 M Ca^{2+} standard solution to the internal solution.

3. To confirm the free $[Ca^{2+}]$, measure 100 mL of each Ca^{2+} calibration solution for Fluo-5N measurements and pour the

solutions into a clean 150 mL beaker. Rinse the electrode with deionized water, blot it dry, and place the electrode into the sample.

4. Stir the solution thoroughly. The concentration of the sample will be indicated (in mV) on the meter.

5. Fill the reservoirs of the superfusion system with relaxing, permeabilization, and calibration solutions and remove any air bubble from the tubing. Adjust the speed of superfusion to approximately 3 mL/min.

6. Stimulate the cells in NT solution at 1 Hz for 2 min. Switch off the stimulation to record resting [Ca^{2+}]. We define resting [Ca^{2+}] as [Ca^{2+}] 1 min after cessation of stimulation.

7. For permeabilization, expose cardiomyocytes for 5 min to relaxing solution and then 20 s to permeabilization solution [16].

8. After permeabilization, let the cells reach the equilibrium in each calibration solution for 20 s. During this period, continuously record changes in fluorescence. Minimal Fluo-5 (F_{min}) is measured during exposure to Ca^{2+}-free calibration solution with 10 mM caffeine and maximal Fluo-fluorescence (F_{max}) during exposure to calibration solution containing a saturating free [Ca^{2+}] of 10 mM (see **Note 21**).

9. Plot the nuclear envelope Fluo-5N fluorescence versus [Ca^{2+}] and fit the concentration-response curves using the Hill equation: $F = \{(F_{max} - F_{min})/(1 + (K_d/[Ca^{2+}])^n)\} + F_{min}$.

10. Use the constructed curve to determine the in situ Ca^{2+} dissociation constant (K_d) and the Hill coefficient (n) (see **Note 19**).

11. To transform Fluo-5N fluorescence into free [Ca^{2+}] use the equation [Ca^{2+}] = K_d $(F - F_{min})/(F_{max} - F)$, where F is the fluorescence intensity at any given time (see **Note 20**).

4 Notes

1. Plastic labware must be used for all low-level calcium measurements. In case you use any glass container, wash it in a labware dishwasher using the analytical grade washing and drying program. Rinse it three times with Milli-Q water, two times with 5 mM EGTA solution (pH 7.4), and three times with ultrapure deionized water (see **Note 2**).

2. For low-level calcium measurements (lower than 10^{-5} M or 0.4 ppm) post-processing of Milli-Q water is necessary to move from the sub-ppb to the sub-ppt Ca^{2+} contamination levels. It can be achieved by adding a purification unit specialized for trace element analysis, such as the Q-POD Element Unit (Merck Millipore, Darmstadt, Germany) to the regular Milli-Q system or by processing Milli-Q water with Chelex 100 resin (Chelex 100, 200–400 mesh sodium, Bio-Rad, CA, USA) [17].

3. The Ca^{2+} concentration of the samples is determined by comparison to the standards and it is strongly affected by the ionic strength of the solution. For high-ionic-strength samples (i.e., ionic strength of 0.1 M or greater), prepare standards with a background composition similar to that of the samples. For samples that have ionic strength less than 0.1 M, commercially available Calcium Ionic Strength Adjuster (ISA, Thermo Scientific, Beverly, MA, USA) can be used to provide a constant background ionic strength for samples and standards.

4. Filtration with Büchner funnel can also be used to more rapidly prepare the resin.

5. Mixing 10 mL of the 0.1 M standard solution with 90 mL of background standard solution #2 will dilute the concentration of background standard solution #2 components by 10 %. As having constant ionic strength of the solution is crucial for accurate measurements, prepare 100 mL of "concentrated" background standard solution #2 that will give the desired concentration when diluted by 10 % and use it for preparing the first Ca^{2+} standard solution.

6. If you need to adjust the final dye concentration for your experimental model, consider that the final concentration of DMSO in the cell medium should be ≤ 0.1 %, which on its own showed no effect on cardiomyocyte morphology and contractile function [18].

7. Prepare the stock solution with a final concentration of 100 mM EGTA. To have as accurate EGTA concentration as possible [15], we suggest to bake the EGTA at 150 °C for 3 h before dissolving. While slowly adding 38.035 g of EGTA to the 1 L of ultrapure deionized water, stir the solution and monitor the pH continuously and when it drops below 7.0, bring it back with KOH to about 7.4 to allow further EGTA to dissolve.

8. The calibration solutions contain 2,3-butanedione monoxime (BDM) to prevent contracture of the cardiomyocytes at high $[Ca^{2+}]$. Alternatively, 10 µM blebbistatin may well be used.

9. Ca^{2+} ionophore A23187 may aggregate over time in aqueous systems. You should prepare and use solutions on the same day. A23187 is soluble in DMSO with heating or sonication. Prepare a 100 mM stock solution in anhydrous DMSO and protect it from light. Add microliter amounts of this solution to your aqueous system with rapid mixing. During the course of experiments, always check for signs of precipitation.

10. Rotenone is a metabolic inhibitor, added to block active Ca^{2+} transport systems and allow equilibration of $[Ca^{2+}]$ between the extracellular medium and the cell interior. It is soluble in organic solvents such as DMSO, absolute ethanol, or dimethyl

formamide (DMF). However, upon exposure to light and air, rotenone in organic solvents decomposes and is oxidized. The solution, previously transparent, becomes brownish. It is imperative to protect the stock solution from direct light using aluminum foil and to prepare rotenone solution on the day of the experiment.

11. When we add CaCl$_2$ to the solution it will dissociate to Ca^{2+} plus two Cl$^-$. Ca^{2+} will bind to EGTA, displacing two protons (H$^+$). The release of two H$^+$ will make the solution more acidic. While slowly adding CaCl$_2$ to the 1 EGTA solution, stir the solution and monitor and adjust the pH continuously. Rotenone degradation and A23187 aggregation have also been found to be dependent on the pH of the aqueous system in which it is dissolved. Make sure to add these substances when final pH adjustment has been made, just before the measurements.

12. Slowly thaw the laminin product at 2–8 °C to avoid the formation of a gel that cannot be reactivated for use. Dilutions of final laminin solution used for coating may vary, but routinely fall in the range of 5–100 mg/mL. To reduce the coating time, you can incubate cover slips at 37 °C for 1–2 h. Laminin-coated cover slips may be stored for ~1 month at 2–8 °C.

13. We suggest using some of the previously described standard enzyme-based Langendorff perfusion protocols for isolating cardiomyocytes from mouse, rat [4], or rabbit [19] heart.

14. Due to the activity of the organic anion transporters, dye may be exported from the cytoplasm to the extracellular medium or actively sequestrated into cellular organelles. If a reduction in Fluo-4 signal or its obvious accumulation in subcellular structures such as mitochondria over time is observed, 250 µM sulfinpyrazone or 1 mM probenecid, inhibitors of organic anion transporters, may be included in the calibration solutions. Stock solutions of sulfinpyrazone and probenecid are necessarily quite alkaline; it is therefore important to readjust the pH of media to which they have been added. Fluo-5N compartmentalization into SR, ER, and NE depends on esterase activity in the compartment which can vary among cell types. For example, Fluo-5N works well in rabbit, but not in rat or mouse cardiomyocytes.

15. Adequate time must be allowed for electrode stabilization. Longer response time will be needed at low-level measurements. Stir all standards and samples at a uniform rate.

16. Samples and standards should be at the same temperature, since electrode potentials and EGTA affinity are dramatically influenced by changes in temperature. A 1 °C difference in temperature can result in substantial errors in the inferred [Ca^{2+}].

17. MaxChelator is an experimental program using various constants and assumptions for its operation. No program is as good as a calibrated electrode. The equations cannot take into account every consideration and therefore, when measuring low amounts of free Ca^{2+} in a buffer system, always confirm the actual values by ion-selective electrode for Ca^{2+}. More detailed theoretical information behind the MaxChelator can be found in [15].

18. The dissociation constant of Ca^{2+}-EGTA buffer system is a function of pH, temperature (*see* **Note 16**), and ionic strength (*see* **Note 3**). For example, in an EGTA buffer (total EGTA 2 mM, total Ca^{2+} 1 mM, ionic strength 0.1 M at 37 °C), changing the pH from 7.4 to 7.0 can result in the ionized calcium increasing by more than 200 nM, a change that is approximately twice the magnitude of that found in resting cells [4]. This requires the pH to be maintained within ±0.01 U in all calibration solutions.

19. As Fluo-4 and Fluo-5 are stoichiometric Ca^{2+} buffers that bind one mole of Ca^{2+} per one mole of indicator with hyperbolic saturation, fit of the Ca^{2+}-dependent change in fluorescence should give the value of Hill coefficient (n) around 1.0. Error in Hill coefficient can indicate inappropriate control of $[Ca^{2+}]$.

20. There are several approximations of the Grynkiewicz equation based on measuring the resting fluorescence of intact cells. In this way, the transformation of changes in fluorescent signal into absolute $[Ca^{2+}]$ may be easier in practice, as these equations do not require measurements of F_{min} and F_{max} for each cell studied. We suggest to *see* refs. 4, 18 for more details.

21. In cardiomyocytes, F_{max} for $[Ca^{2+}]_{SR}$ can be confirmed by exposing of an intact myocyte to 1 µmol/L isoproterenol, then 0.5 mmol/L tetracaine to block SR Ca^{2+} leak, and finally to 0 Na^+ solution to drive $[Ca^{2+}]_{cyto}$, $[Ca^{2+}]_{SR}$, and $[Ca^{2+}]_{NE}$ up [16]. However, measurements in other cell types should account for specific Ca^{2+} channels and pumps that can induce SR and NE Ca^{2+} depletion and Ca^{2+} loading.

References

1. Bers DM (2008) Calcium cycling and signaling in cardiac myocytes. Annu Rev Physiol 70:23–49

2. Zhang SJ, Zou M, Lu L et al (2009) Nuclear calcium signaling controls expression of a large gene pool: identification of a gene program for acquired neuroprotection induced by synaptic activity. PLoS Genet 08:e1000604

3. Kockskamper J, Seidlmayer L, Walther S et al (2008) Endothelin-1 enhances nuclear Ca^{2+} transients in atrial myocytes through ins(1,4,5)

P_3-dependent Ca^{2+} release from perinuclear Ca^{2+} stores. J Cell Sci 121:186–195

4. Ljubojevic S, Walther S, Asgarzoei M et al (2011) In situ calibration of nucleoplasmic versus cytoplasmic Ca^{2+} concentration in adult cardiomyocytes. Biophys J 100: 2356–2366

5. Wu X, Zhang T, Bossuyt J et al (2006) Local $InsP_3$-dependent perinuclear Ca^{2+} signaling in cardiac myocyte excitation-transcription coupling. J Clin Invest 116:675–682

6. Bootman MD, Thomas D, Tovey SC et al (2000) Nuclear calcium signalling. Cell Mol Life Sci 57:371–378

7. Gerasimenko O, Gerasimenko J (2004) New aspects of nuclear calcium signalling. J Cell Sci 117:3087–3094

8. Echevarria W, Leite MF, Guerra MT et al (2003) Regulation of calcium signals in the nucleus by a nucleoplasmic reticulum. Nat Cell Biol 5:440–446

9. Gee KR, Brown KA, Chen WN et al (2000) Chemical and physiological characterization of fluo-4 Ca^{2+}-indicator dyes. Cell Calcium 27: 97–106

10. Thomas D, Tovey SC, Collins TJ et al (2000) A comparison of fluorescent Ca^{2+} indicator properties and their use in measuring elementary and global Ca^{2+} signals. Cell Calcium 28: 213–223

11. Connor JA (1993) Intracellular calcium mobilization by inositol 1,4,5-trisphosphate: Intracellular movements and compartmentalization. Cell Calcium 14:185–200

12. Giovannardi S, Peres A (1997) Nuclear and cytosolic calcium levels in NIH 3T3 fibroblasts. Exp Biol 2:1–9

13. McDonough PM, Button DC (1989) Measurement of cytoplasmic calcium concentration in cell suspensions: correction for extracellular fura-2 through use of Mn^{2+} and probenecid. Cell Calcium 10:171–180

14. Mitsui M, Abe A, Tajimi M et al (1993) Leakage of the fluorescent Ca^{2+} indicator fura-2 in smooth muscle. Jpn J Pharmacol 61:165–170

15. Bers DM, Patton CW, Nuccitelli R (1994) A practical guide to the preparation of Ca^{2+} buffers. Methods Cell Biol 40:3–29

16. Shannon TR, Guo T, Bers DM (2003) Ca^{2+} scraps: local depletions of free [Ca^{2+}] in cardiac sarcoplasmic reticulum during contractions leave substantial Ca^{2+} reserve. Circ Res 93:40–45

17. Thulin E (2002) Purification of recombinant calbindin D9k. Methods Mol Biol 172:175–184

18. Cannell MB, Cheng H, Lederer WJ (1994) Spatial non-uniformities in [Ca^{2+}]$_i$ during excitation-contraction coupling in cardiac myocytes. Biophys J 67:1942–1956

19. Pogwizd SM, Qi M, Yuan W (1999) Upregulation of Na$^+$/Ca^{2+} exchanger expression and function in an arrhythmogenic rabbit model of heart failure. Circ Res 85: 1009–1019

Chapter 13

Assessing GPCR and G Protein Signaling to the Nucleus in Live Cells Using Fluorescent Biosensors

Julie Bossuyt and Donald M. Bers

Abstract

G protein-coupled receptor (GPCR) signaling cascades regulate a wide variety of cellular processes and feature prominently in many cardiovascular pathologies. As such they represent major drug targets and discovering novel aspects of GPCR signaling provide important opportunities to identify additional potential therapeutic approaches to reverse or prevent cardiac remodeling and failure. Monitoring cellular trafficking of signaling components and specific protein kinase activities using fluorescent biosensors has provided key insight into stress/GPCR-induced kinase signaling networks and their effect on cardiac gene expression. Herein we describe the protocols for the expression, visualization (by confocal microscopy), and interpretation of data obtained with such biosensors expressed in adult cardiomyocytes. Our focus is on the cellular trafficking of class II histone deacetylases (i.e., HDAC5) and on the FRET sensor (Camui) for calmodulin-dependent protein kinase II (CaMKII).

Key words Fluorescence resonance energy transfer, Biosensor, Fluorescence imaging, Histone deacetylase, Calmodulin-dependent protein kinase II

1 Introduction

G protein coupled receptors (GPCRs) control and influence a wide array of cellular functions through the initiation of complex signaling cascades with both acute (e.g., calcium handling) and chronic effects (e.g., gene expression). Among the most investigated and clinically targeted cardiac GPCRs include the α- and β-adrenergic and angiotensin receptors but there are well over 400 GPCRs (including endothelin, dopamine, opioid, muscarinic) [1]. Profound neurohumoral activation and dysregulation of β-adrenergic signaling are considered hallmarks of heart failure. The heightened and continuous GPCR stimulation is initially protective but ultimately leads to molecular, cellular, and interstitial changes that worsen cardiac function [2]. This cardiac remodeling process is associated with alterations in gene expression [3]. The nuclear envelope presents a major access barrier to the target genes. So determining how

Bruce G. Allen and Terence E. Hébert (eds.), *Nuclear G-Protein Coupled Receptors*, Methods in Molecular Biology, vol. 1234, DOI 10.1007/978-1-4939-1755-6_13, © Springer Science+Business Media New York 2015

the nuclear translocation of signal transducers, such as protein kinases and transcriptional regulators, is accomplished in a stimulus-induced manner and is key to our understanding of the relevant signal-transduction pathways.

Class II histone deacetylases (HDACs) are key signal-responsive repressors of transcription, favoring a condensed chromatin state [3, 4]. They lack intrinsic DNA-binding activity but exert their effects via direct association with transcription factors (e.g., MEF2) or indirectly via bridging cofactors [5]. The subcellular localization of class II HDACs, and hence their regulation of gene transcription, is finely regulated by diverse posttranslational modifications [6]. Numerous studies have shown that calmodulin-dependent protein kinase II (CaMKII)- or protein kinase D (PKD)-dependent phosphorylation of HDAC triggers its nuclear export and subsequent transcriptional activation [7–12]. In contrast, phosphorylation by other protein kinases, such as PKA, induces nuclear import of HDACs [13, 14]. Other posttranslational modifications of HDACs that modulate their subcellular localization include oxidation and sumoylation [15, 16]. Thus, fluorescence measurements of the nucleocytoplasmic shuttling of HDACs, and of HDAC mutants lacking specific regulatory sites, provide unique insights into regulatory effects on transcription by specific GPCR signals.

How HDAC kinases such as CaMKII and PKD translate stimuli into downstream functional effects is also of particular interest, but the dynamic modulation of these kinases in intact cardiomyocytes is still largely unknown. Hayashi developed Camui, a FRET-based biosensor consisting of full-length CaMKII sandwiched between donor and acceptor fluorophores [17]. This construct produces a robust FRET signal when CaMKII is in the autoinhibited state, but is reduced upon activation of the kinase (Fig. 1). Expression in cardiomyocytes (and neurons) permits monitoring of the spatial and temporal dynamics of CaMKII activity [18]. Since this method is non-disruptive, it offers several advantages over more traditional approaches to measuring kinase activity (e.g., snapshots of CaMKII activation state via immunoblots probing for posttranslational modifications of CaMKII or CaMKII target phosphorylation). Moreover, the specific role of posttranslational modifications of CaMKII can be examined by using mutant forms of Camui lacking the target sites for particular modifications [18, 19].

2 Materials

2.1 Cardiomyocyte Culture

1. PC-1 media supplemented with 5 % penicillin/streptomycin.

2. Natural mouse laminin (Life Technologies) diluted to 0.2 mg/mL in buffer containing 10 mM Tris–HCl (pH 7.2) and 1 mM EGTA.

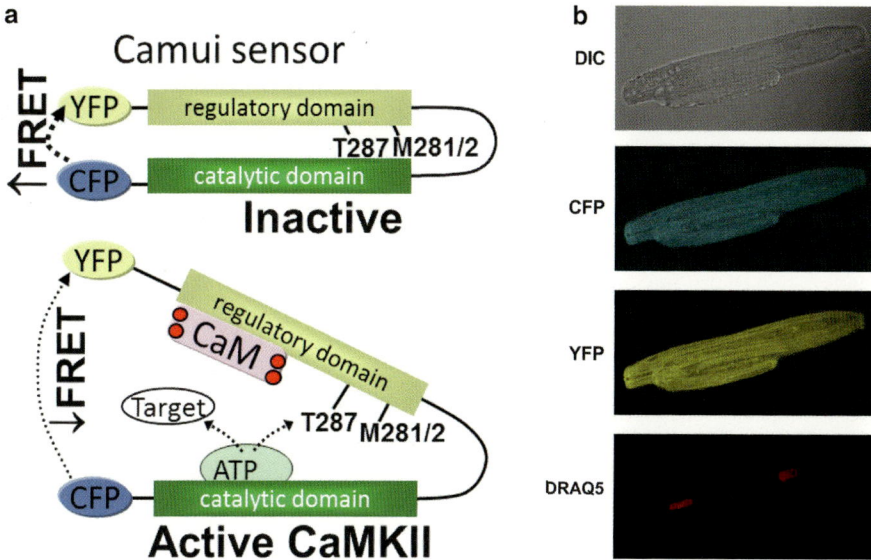

Fig. 1 *Camui: A FRET biosensor for CaMKII.* (**a**) Camui consists of the full-length kinase sandwiched between the CFP/YFP FRET pair. Binding of Ca-CaM triggers a conformational change in the kinase, which increases the distance between the catalytic and regulatory domains effectively reducing FRET between the CFP/YFP FRET pair. (**b**) Representative images of a cardiomyocyte expressing Camui where the nuclei are counterstained with DRAQ5

3. Uncoated 2-well chambered coverglass (Nunc Lab-Tek) or sterile 25 mm round coverglass #1.5 thickness.

4. Purified adenovirus encoding HDAC, Camui, or mutated versions of these biosensors.

5. 8-Well Nunclon rectangular culture dish.

6. C-PACE EP culture pacer (Ionoptix) with electrode assembly for 8-well Nunclon rectangular culture dish.

7. Standard sterile cell culture pipettes and laminar flow cabinet to maintain sterility as well as 37 °C water bath and CO_2 incubator.

2.2 Cardiomyocyte Stimulation and Data Acquisition

1. Tyrode's solution (140 mM NaCl, 4 mM KCl, 1 mM $MgCl_2$, 1.8 mM $CaCl_2$, 10 mM HEPES, 10 mM glucose, pH 7.4).

2. Warner Quick Exchange heated platform and imaging chamber with field stimulation (optional: platform temperature controlled by TC 324B with flow and heating of perfusate controlled by eight-channel perfusion valve control system (VC-8) and in-line solution heater (SHM-8)). Field stimulation was controlled with a Grass stimulator (e.g., S88X).

3. *Confocal laser scanning microscope*: The experimental procedure is described for an Olympus FV1000 laser scanning confocal system with spectral detection on an Olympus IX81 inverted

microscope (*see* **Note 1**). This system is also equipped with a motorized stage permitting multi-area time-lapse measurements, a focal drift compensation system (ZDC), and the following laser lines: 405, 440, 488, 515, 559, and 635 nm. However, many confocal microscopes would suffice, as long as the necessary laser lines (see below) are available. Image analysis can be done with the confocal software or post-acquisition with Image J (http://imagej.nih.gov/ij/).

4. 60x oil-immersion objective lens (PLAPO60xOI, NA 1.40).

5. *Drugs and cellular markers*: For the identification of the sarcolemma and the nucleus in live cells, wheat germ agglutinin Alexa Fluor 647 conjugate (Life Technologies) and DRAQ5 (Cell Signaling) were used, respectively. Drugs are made up fresh or kept frozen at a 1,000× stock concentration in small aliquots to avoid frequent freeze-thaws.

3   Methods

The following section describes a standard protocol for the culture of adult cardiomyocytes with adenovirus-mediated expression of recombinant protein, followed by detailed methods for confocal measurements of HDAC5-GFP translocation and FRET measurements of CaMKII activation using the Camui biosensor. However, these specific protocols are easily adaptable for the imaging of a wide variety of fluorescently tagged proteins and similar biosensors and can be applied to other cell types.

3.1 Cardiomyocyte Culture and Biosensor Expression

1. Clean glass cover slips by washing with 1 M HCl solution followed by multiple rinses in double-distilled water. Make sure to individually separate the cover slips since they tend to stick together. Then dry the cover slips on a clean surface before autoclaving.

2. Place the sterile cover slips in 8-well culture dishes and coat them with laminin for 5 min.

3. Remove the laminin solution and let cover slips dry while you resuspend and then gravity settle the freshly isolated cardiomyocytes in the culture media, in conical tubes. Several gravity settlings may be needed to remove dead cells if the isolation procedure was suboptimal.

4. Resuspend the cardiomyocyte pellet in fresh culture media, seed cardiomyocytes onto the cover slips at a density of $0.5–1 \times 10^3/$ cm^2, and let attach for 30–60 min in the incubator.

5. Gently replace the culture media to remove non-adherent cells and add the appropriate adenovirus at an MOI of 10–500 (*see* **Note 2**). Place the culture dish in the C-Pace dish (i.e., the

electrode assembly) and select applicable pacing protocol (we typically use 35 V, 0.2 Hz, 5 ms). After 2 h replace the culture media and resume culture with pacing overnight (*see* **Note 1**).

6. Replace the culture media the next day.

7. For experiments, transfer a cover slip to the chamber holder and cover with Tyrode's solution; avoid fluid leaks.

8. Remove any remaining fluid from the bottom of the cover slip.

9. Add a small drop of immersion oil to the objective before placing the holder with the cover slip on the microscope stage.

10. Carefully adjust perfusion of the chamber if performing pacing experiments. For agonist-driven experiments not requiring constant perfusion, the experiment is run with a static bath (of known volume) and the reagent (prediluted from a concentrated stock solution) is carefully applied into the bath by pipette. Movement artifacts may occur whilst making the additions, but with care these can be minimized.

3.2 Confocal Imaging of GFP-HDAC5

1. Turn on the confocal system and the lasers (*see* **Note 2**). The laser for GFP excitation is 488 nm and for DRAQ5 is 635 nm. Emission can be 500–600 nm for GFP and LP > 650 for DRAQ5. The images can be obtained simultaneously for GFP and DRAQ5 (Fig. 2).

2. Focus the cells with normal light and evaluate overall cardiomyocyte quality before adding the DRAQ5 (0.5–1 μl/1 mL Tyrode's). After 5-min incubation with DRAQ5 the experiment can commence.

3. Acquire a pre-activated image. To identify detailed structure of the cell of interest, collect both a differential interference contrast (DIC) image and fluorescent image using the appropriate control settings: frame scan with 1,024 × 1,024 pixel frame size, scan speed 8–10 μs/pixel, 12 bit imaging, and pinhole set with 1 airy unit. The detector gain, laser strength, and amplifier offset should be determined empirically in order to obtain a fluorescent image with strong signal/noise ratio and without pixel saturation. The minimal laser intensity needed for clear signals is beneficial with respect to minimizing photobleaching. Several images can be captured to establish a baseline for time-lapse experiments (*see* **Notes 3–6**).

4. Apply the drug or pacing protocol and then initiate time-lapse imaging (in the pacing experiments the images were taken after cessation of stimulation). For nucleocytoplasmic shuttling, time intervals of 10 min for a period of 1–2 h are sufficient. Watch for focal drift during the time-lapse and correct if necessary (*see* **Note 7**).

5. Image analysis with Image J: The DRAQ5 signal is used to define the nuclear region of interest (ROI) so that the mean

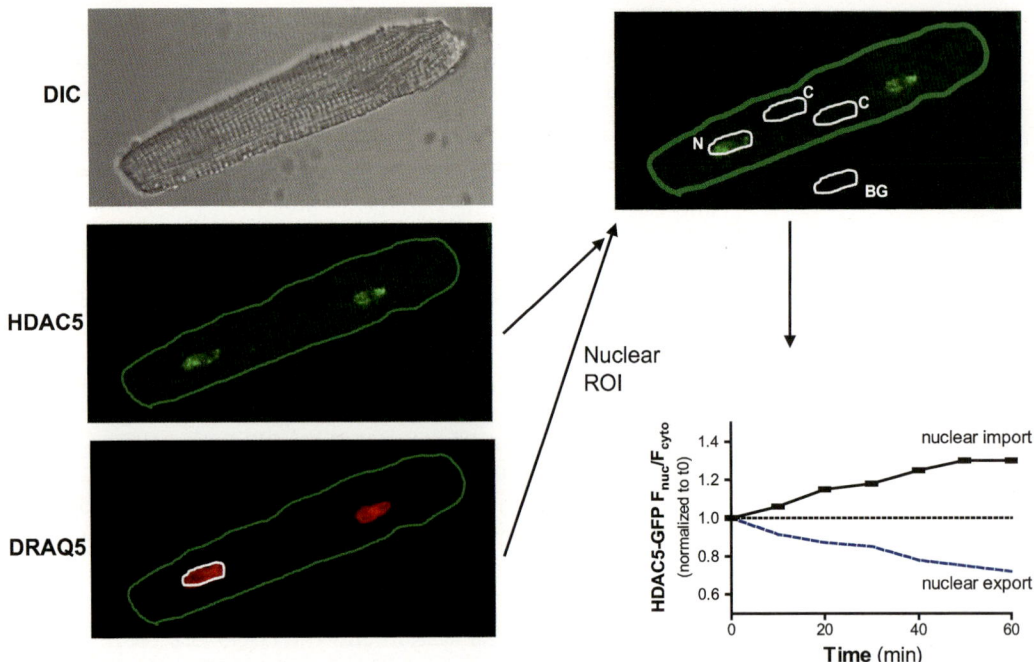

Fig. 2 *Analysis of GFP-HDAC5 translocation in cardiomyocytes.* Representative confocal images are shown on left (*top*: DIC image, *middle*: GFP-HDAC5, *bottom*: DRAQ5 nuclear stain). To obtain the F_{nuc}/F_{cyto} ratio, the GFP fluorescence signal in the DRAQ5-defined nuclear ROI was divided by the average of three such ROIs randomly spaced throughout the cytosol (all signals were also corrected for background signals). A F_{nuc}/F_{cyto} ratio larger than 1 indicates net nuclear accumulation of HDAC5, whereas a ratio smaller than 1 indicates nuclear export

(or integrated) fluorescence value for the area enclosed by the ROI (F_{nuc}) can be calculated for each GFP image of the experiment. For the mean (or integrated) fluorescence value of the cytosol (F_{cyto}), the "nuclear ROI" is placed in 2–3 different areas of the cytosol and averaged. These values are corrected for the background signal from areas in the image lacking cells. The data can then be plotted as raw or normalized fluorescence values against time to indicate the extent of nucleocytoplasmic shuttling. The F_{nuc}/F_{cyto} ratio of 1 represents even distribution of the proteins between the nucleus and cytoplasm (*see* **Note 7**).

3.3 FRET Measurements of Camui

1. The appropriate lasers to turn on are 440 nm for Camui and 635 nm for WGA or DRAQ5 stains (*see* **Note 1**). The Camui and marker images are taken sequentially with the emissions for the markers at LP 650 nm and for Camui at 460–500 nm for CFP and 515–615 nm for YFP so that a FRET ratio can be calculated (Fig. 2).

2. Focus the cells with normal light and evaluate overall cardiomyocyte quality before adding the subcellular compartment

marker (0.5–1 μL/1 mL Tyrode's). After a 5-min incubation, the experiment can commence. When selecting a cardiomyocyte to begin recording, preference should be given to those cells with only a moderate expression of Camui. Excessive expression of the sensor may result in nonphysiological localization and or function. Moreover, endogenous calmodulin can be limiting for Camui studies, and that potential problem is most prominent when Camui is expressed at high levels (*see* **Notes 2, 6, 8** and **9**).

3. Acquire a baseline. To identify detailed structure of the cell of interest, collect both a DIC image and fluorescent image using the appropriate control settings: frame scan with $1,024 \times 1,024$ pixel frame size, scan speed 8–10 μs/pixel, 12 bit imaging, and pinhole set with 1 airy unit. The detector gain, laser strength, and amplifier offset should be determined empirically in order to obtain a fluorescent image with strong signal/noise ratio and without pixel saturation (*see* **Note 7**). Several images can be captured to establish a baseline for time-lapse experiments. If the intent is to compare baseline FRET ratios, great care should be taken to maintain the same image acquisition settings for all the experiments. If that is not the case, then one can only reasonably compare the changes in FRET during a given stimulus as a readout of changes in activity.

4. Apply the drug or pacing protocol, and then initiate time-lapse imaging. For nucleocytoplasmic shuttling time intervals of 1–10 min for a period of 30–60 min are sufficient. Watch for focal drift during the time lapse and correct if necessary. When using the continual perfusion system, note the timings of the additions made and how long the cells were exposed to the drug(s) (*see* **Notes 10** and **11**).

5. Image analysis: If compartment-specific stains were used, those images can be thresholded and skeletonized to provide the compartment-specific ROI. For each ROI the average fluorescence intensities for donor (F_{CFP}) and acceptor (F_{YFP}) emission are quantified and corrected for background fluorescence in an area of the image without the cell. The FRET ratio (F_{CFP}/F_{YFP}) for each ROI selected is plotted over time.

6. Alternative protocol (acceptor photobleach method): The nondestructive ratiometric FRET approach described above is well suited for monitoring time-dependent FRET changes but quantification of the FRET efficiency is more appropriate with an acceptor photobleach approach. Here FRET efficiency is measured as the increase in donor (CFP) fluorescence upon photobleach of YFP (which eliminates energy transfer from CFP to YFP). So in its most basic form, this experiment will acquire a pre- and postbleach image with the postbleach image acquired as quickly as possible after acceptor photobleach. The laser settings for photobleaching must be determined

empirically (F_{YFP} signal with excitation 514 nm reduced by >95 %), but usually the process consists of 50–100 continual laser iterations in the ROI with >80 % transmission power. This can be done in the same cell at the end of a time-lapse experiment or as a separate experiment.

4   Notes

1. Comments on cardiomyocyte culture. Pacing of cardiomyocytes during short-term culture is done to preserve both the T-tubular structure and major functional characteristics such as action potentials, calcium handling, and contractility. Alternatively 500 nM cytochalasin D, a concentration too low to have mechanical uncoupling effects, was shown to also preserve cardiomyocyte morphology and function [22]. The optimal culture media for cardiomyocytes of different species differs (e.g., rat myocytes prefer supplemented M199 media). Adenovirus incubation periods are typically only 2 h but with low-expressing constructs it may help to extend the exposure period to overnight.

2. Clear resolution of the localization of either biosensor requires confocal microscopy, although the Camui FRET measurements can also be performed without subcellular spatial resolution using an epifluorescence microscope. Ideally this system would be equipped with simultaneous, dual-emission wavelength detection.

3. Overexpression of any protein may influence cardiomyocyte behavior. It is therefore advisable to ensure that fluorescent biosensors are expressed at appropriate "tracer" levels. Moreover the correct targeting of the biosensors should be verified using immunocytochemistry of the native endogenous protein whenever possible.

4. When setting up the imaging parameters, the aim should be to maximize image quality while minimizing the potential for photobleaching or saturation of the response and also permitting an appropriate sampling rate. In our hands, the HDAC5 translocation in adult cardiomyocytes is <50 % and the Camui response <20 %, so it is advisable to set the basal fluorescence level to <50 % of the maximum detectable fluorescence level. In order to maximize the image quality/signal size we also prefer to use the gain settings on the PMT rather than the laser strength to avoid photobleaching. It may be necessary to sacrifice some image quality for scanning speed if photobleaching becomes a problem.

5. Photobleach evaluation/corrections: Time-sequential images of untreated cardiomyocytes expressing the biosensor are used to evaluate the temporal evolution of the average fluorescent

intensities. The decline in signal over the duration of the experiment should be kept minimal (~1 %) but the data can be corrected for significant photobleaching as follows: an averaged data set of the temporal evolution of the fluorescence intensity is normalized before fitting with a nonlinear least-square regression with an exponential model to generate a photobleaching curve. The photobleach-corrected data is obtained by dividing the uncorrected fluorescence value at a given time point by the one extracted from the curve. Note that this correction is redundant in the HDAC5 experiments if you express the fluorescence as F_{Nuc}/F_{Cyto} because the percent photobleach should be the same for both compartments.

6. Susceptibility to artifacts. These are less common with the HDAC5 probe but there is a propensity of GFP-HDAC5 to aggregate in the cytosol with high levels of overexpression and/or poor cell quality. The FRET sensors are more susceptible to artifacts, which can be due to contaminating autofluorescence signals or environmental effects on the fluorophores (YFP especially is sensitive to pH changes). A genuine change in CFP-YFP FRET should be reflected in both the individual donor and acceptor traces (e.g., not only a decrease in YFP emission but also a small increase in CFP emission).

7. For the nucleocytoplasmic shuttling analysis, it is prudent to err on the conservative side when selecting the nuclear ROI. The potential problem here is that fluorescence from above or below the plane in which the nucleus is widest will be influenced by perinuclear fluorescence. This will be most important when perinuclear signals are substantial and or when one wants to differentiate between nuclear, perinuclear, and bulk cytosolic fluorescence.

8. When using the Camui sensor for the first time, it is prudent to verify that the image acquisition settings do not elicit significant autofluorescence contributions by checking the signal obtained in non-transfected cells. It is also advisable to determine the dynamic range of the FRET sensor in your system by measuring the baseline FRET ratio in the presence of KN93 (minimal activity, maximal FRET) and ionomycin/1 mM Ca/10 μM cytochalasin D (for maximal activity, minimal FRET).

9. For the Camui experiments, it is important to note that CaM signals may be limiting especially in cells with high Camui expression. This is partly due to the decline in CaM expression levels during culture, which can be minimized by optimizing the culture conditions but could also be countered by coexpression of CaM with Camui [20, 21].

10. We typically perform these experiments in ambient temperature but improved kinetics of signaling could be observed at 37 °C.

11. Many agonists and inhibitors have an intrinsic fluorescence, which may interfere with the biosensor signals. So any drug is best tested for its contaminating fluorescence in both the background and non-biosensor-expressing cells.

Acknowledgements

This work was supported by NIH grants P01 HL080101 and R37 HL30077 (DMB) and R01 HL103933 (J.B.).

References

1. Salazar NC, Chen J, Rockman HA (2007) Cardiac GPCRs: GPCR signaling in healthy and failing hearts. Biochim Biophys Acta 1768: 1006–1018

2. Cohn JN, Ferrari R, Sharpe N (2000) Cardiac remodeling–concepts and clinical implications: a consensus paper from an international forum on cardiac remodeling. Behalf of an International Forum on Cardiac Remodeling. J Am Coll Cardiol 35:569–582

3. Backs J, Olson EN (2006) Control of cardiac growth by histone acetylation/deacetylation. Circ Res 98:15–24

4. McKinsey TA (2007) Derepression of pathological cardiac genes by members of the CaM kinase superfamily. Cardiovasc Res 73: 667–677

5. Haberland M, Montgomery RL, Olson EN (2009) The many roles of histone deacetylases in development and physiology: implications for disease and therapy. Nat Rev Genet 10: 32–42

6. Eom GH, Kook H (2014) Posttranslational modifications of histone deacetylases: implications for cardiovascular diseases. Pharmacol Ther 143:168–180. doi:10.1016/j.pharmthera.2014.02.012

7. McKinsey TA, Zhang CL, Lu J et al (2000) Signal-dependent nuclear export of a histone deacetylase regulates muscle differentiation. Nature 408:106–111

8. McKinsey TA, Zhang CL, Olson EN (2000) Activation of the myocyte enhancer factor-2 transcription factor by calcium/calmodulin-dependent protein kinase-stimulated binding of 14-3-3 to histone deacetylase 5. Proc Natl Acad Sci U S A 97:14400–14405

9. Zhang CL, McKinsey TA, Chang S et al (2002) Class II histone deacetylases act as signal-responsive repressors of cardiac hypertrophy. Cell 110:479–488

10. Vega RB, Harrison BC, Meadows E et al (2004) Protein kinases C and D mediate agonist-dependent cardiac hypertrophy through nuclear export of histone deacetylase 5. Mol Cell Biol 24:8374–8385

11. Wu X, Zhang T, Bossuyt J et al (2006) Local InsP$_3$-dependent perinuclear Ca^{2+} signaling in cardiac myocyte excitation-transcription coupling. J Clin Invest 116:675–682

12. Bossuyt J, Helmstadter K, Wu X et al (2008) Ca^{2+}/calmodulin-dependent protein kinase IIδ and protein kinase D overexpression reinforce the histone deacetylase 5 redistribution in heart failure. Circ Res 102:695–702

13. Ha CH, Kim JY, Zhao J et al (2010) PKA phosphorylates histone deacetylase 5 and prevents its nuclear export, leading to the inhibition of gene transcription and cardiomyocyte hypertrophy. Proc Natl Acad Sci U S A 107: 15467–15472

14. Chang CW, Lee L, Yu D et al (2013) Acute β-adrenergic activation triggers nuclear import of histone deacetylase 5 and delays G$_q$-induced transcriptional activation. J Biol Chem 288: 192–204

15. Haworth RS, Stathopoulou K, Candasamy AJ et al (2012) Neurohormonal regulation of cardiac histone deacetylase 5 nuclear localization by phosphorylation-dependent and phosphorylation-independent mechanisms. Circ Res 110:1585–1595

16. Kirsh O, Seeler JS, Pichler A et al (2002) The SUMO E3 ligase RanBP2 promotes modification of the HDAC4 deacetylase. EMBO J 21:2682–2691

17. Takao K, Okamoto K, Nakagawa T et al (2005) Visualization of synaptic Ca^{2+} /calmodulin-dependent protein kinase II activity in living neurons. J Neurosci 25:3107–3112

18. Erickson JR, Patel R, Ferguson A et al (2011) Fluorescence resonance energy transfer-based

sensor Camui provides new insight into mechanisms of calcium/calmodulin-dependent protein kinase II activation in intact cardiomyocytes. Circ Res 109:729–738

19. Erickson JR, Pereira L, Wang L et al (2013) Diabetic hyperglycaemia activates CaMKII and arrhythmias by O-linked glycosylation. Nature 502:372–376

20. Song Q, Saucerman JJ, Bossuyt J et al (2008) Differential integration of Ca^{2+}-calmodulin signal in intact ventricular myocytes at low and high affinity Ca^{2+}-calmodulin targets. J Biol Chem 283:31531–31540

21. Maier LS, Ziolo MT, Bossuyt J et al (2006) Dynamic changes in free Ca-calmodulin levels in adult cardiac myocytes. J Mol Cell Cardiol 41:451–458

22. Tian Q, Pahlavan S, Oleinikow K et al (2012) Functional and morphological preservation of adult ventricular myocytes in culture by submicromolar cytochalasin D supplement. J Mol Cell Cardiol 52:113–124

Chapter 14

Tandem Affinity Purification to Identify Cytosolic and Nuclear Gβγ-Interacting Proteins

Rhiannon Campden, Darlaine Pétrin, Mélanie Robitaille, Nicolas Audet, Sarah Gora, Stéphane Angers, and Terence E. Hébert

Abstract

It has become clear in recent years that the Gβγ subunits of heterotrimeric proteins serve broad roles in the regulation of cellular activity and interact with many proteins in different subcellular locations including the nucleus. Protein affinity purification is a common method to identify and confirm protein interactions. When used in conjugation with mass spectrometry it can be used to identify novel protein interactions with a given bait protein. The tandem affinity purification (TAP) technique identifies partner proteins bound to tagged protein bait. Combined with protocols to enrich the nuclear fraction of whole cell lysate through sucrose cushions, TAP allows for purification of interacting proteins found specifically in the nucleus. Here we describe the use of the TAP technique on cytosolic and nuclear lysates to identify candidate proteins, through mass spectrometry, that bind to Gβ$_1$ subunits.

Key words Tandem affinity purification, Nuclear isolation, Streptavidin sepharose, Calmodulin sepharose, Agarose beads, Mass spectrometry, Protein isolation, Gβ signaling

1 Introduction

The TAP protocol was first developed by Rigaut et al. for use in yeast, but has since been used for identifying protein interactions in numerous mammalian cell types [1]. The technique was developed to identify multimeric protein complexes under mild, physiological conditions in the context of heterologous expression. Toward this aim, high affinity protein tags are attached to a bait protein, and stable cell lines can be created [2]. As initially developed, TAP used the binding domains *Staphylococcus aureus* Protein A (ProtA) and calmodulin binding domains (CBD), separated by a tobacco etch virus (TEV) protease cleavage site, which resulted in efficient recovery of bait proteins. The tag could be attached to either the N- or C-termini of the bait protein [3]. The bait protein and bound protein interactors are selected through affinity for the IgG ProtA and calmodulin beads and tandem mass spectrometry was

Bruce G. Allen and Terence E. Hébert (eds.), *Nuclear G-Protein Coupled Receptors*, Methods in Molecular Biology, vol. 1234, DOI 10.1007/978-1-4939-1755-6_14, © Springer Science+Business Media New York 2015

initially used to identify proteins in eluates [2]. The technique has since been modified for use in mammalian cell lines such as human embryonic kidney cells (HEK 293) [3, 4].

In addition to the ProtA and CBD tags there are also small peptide tags such as GST (glutathione-S transferase) and MBP (maltose binding protein), small peptide tags such as CBD (choline binding domain), methionine, FLAG, and streptavidin binding domains (SBD), and combination tags such as choline-binding histidine and nanoscale gold particle tags [5–17]. The choice of these tags depends on the protein being isolated, the conditions required for binding and elution of the bait protein, whether or not the tags will disrupt protein binding, and the conditions required for affinity purification [13]. These tags are used with corresponding beads including streptavidin Sepharose beads and magnetic gold particle beads, broadening the repertoire for TAP [5–8]. An overview of these tags and their corresponding affinities has been reported elsewhere [8, 13].

The TAP protocol can also be adapted to engineer half of a TAP tag onto two different proteins, allowing for identification of protein complexes that require two separate proteins, known as bimolecular-TAP [8]. In addition, the TAP protocol can be used for identification of proteins from specific cellular compartments and organelles [18]. Prior to application of the TAP protocol, specific cell fractions of total lysates from cells expressing bait proteins can be isolated by differential centrifugation. Previous work has described the use of TAP with cytoplasmic, nuclear, and chromosomal fractions [18].

The TAP protocol has previously been used to identify protein interactions with the G protein dimer $\beta\gamma$ in whole cell lysates from HEK 293 cells [4]. One of the important advantages of TAP is that eluates can be directly applied for mass spectrometry experiments without having to run 1- or 2D gels to isolate bands of interest. We have also adapted this technique for direct injection of TAP-tagged G protein-coupled receptor (GPCR) eluates, allowing us to understand their signaling networks from the perspective of receptors or G proteins (see below) [1]. Recent evidence shows that G$\beta\gamma$ interacts with a number of different proteins in the nucleus (reviewed in ref. 19). Functional expression of GPCRs, G proteins, and their associated signaling machinery have been detected on the nuclear membranes or in nuclei in various cell types (reviewed elsewhere [19–22]). Several lines of evidence have shown functional interactions of G$\beta\gamma$ in the nucleus with transcription factors. G$\beta_1\gamma_2$ was found to interact with histone deacetylase 5 (HDAC5) to prevent its repression of muscle differentiation factor (MEF2C)-sensitive genes [23]. We showed that G$\beta\gamma$ decreased phorbol 12-myristate 13-acetate (PMA)-stimulated activating protein (AP-1) reporter gene activity in different cell lines [24]. These effects were dependent on G$\beta\gamma$ interactions with members of AP-1 transcription factor complex and HDACs, acting specifically in the nucleus. Gγ_5 can also prevent the repression of adipocyte enhancer binding protein (AEBP1)-sensitive genes by binding to AEBP1 and preventing its interaction with DNA [25]. Gγ$_5$ and AEBP1 are both localized to the nucleus; however the authors did not show whether this interaction occurs specifically in the nucleus.

The extent of such nuclear G protein networks remains to be understood. Functions of distinct Gβγ isoforms in the nucleus remain largely unknown. As an initial screen to better understand nuclear Gβγ signaling networks in the nucleus, we have adapted the TAP protocol for use with nuclear lysates using TAP-tagged Gβ$_1$ subunits. This technique involves isolation of nuclei from HEK 293 cells through a sucrose cushion, lysing nuclei, binding tagged Gβ$_1$ lysates to and eluting them from streptavidin Sepharose beads, followed by binding to and elution from calmodulin Sepharose beads. Samples can then be processed for western blotting or mass spectrometry. Problems encountered in focusing on the nuclear fraction include isolating sufficient protein, contaminating DNA in samples, and defining appropriate buffers for compatibility with mass spectrometry.

Another aspect to consider in defining the function of Gβγ is the number of possible Gβγ isoforms. There are 5 Gβ and 12 Gγ subunits that comprise different dimers expressed in different cell types [19]. For this reason, we have also adopted a split-tag approach, placing half of the TAP tag onto individual Gβ and Gγ subunits. Following tandem affinity purification, this allows identification of proteins that bind to specific Gβγ dimer pairs. These constructs can be used on nuclear isolates as well as on whole cell lysates to identify isoform-specific Gβγ interactions.

The following method describes the use of the TAP tag technique on nuclear lysates followed by liquid chromatography/mass spectroscopy (LC/MS) for protein identification. The technique presents particular challenges in the steps of nuclear isolation, tandem affinity purification, and mass spectrometry, which will be discussed in detail in the text below.

2 Materials

Prepare all solutions using ultrapure Milli-Q water. Particular steps need to be taken to limit contaminants in samples for MS. Contamination can occur from various sources, from skin to plasticware. Any glassware used must be washed thoroughly in Milli-Q water. Keratin contamination in dust is largely from human origin [26]. To reduce this, wear gloves, lab coats, and tie back long hair at all times. Other precautions include ensuring a clean environment, such as working in a safety cabinet or at minimum, carefully cleaning your workspace. Choose plastic containers and pipette tips that are compatible with MS, and avoid using plastic ware to store acids and bases used in the protocol [26]. All plastic ware used must be new. Further contamination can result from autoclaving solutions [26]. If solutions must be sterilized, use 0.22 μm filters.

2.1 Cell Culture

1. Tissue culture dishes (T175).

2. Tissue culture hood.

3. Incubator: 37 °C, 5 % CO_2.

4. Dulbecco's modified Eagle's medium (DMEM) high glucose supplemented with fetal bovine calf serum (FBS) (10 % final concentration) and penicillin/streptomycin (P/S) (100 U/mL final concentration).

5. Puromycin (*see* **Note 1**).

6. Hygromycin.

7. Lipofectamine 2000.

8. TAP-tagged vector pIRESpuro-GLUE with HA-Gβ₁ inserted: human Gβ was cloned by PCR using EcoRI and BamHI into the pIRESpuro-GLUE-N1 vector, described elsewhere [1, 9]. Split-TAP tagging of Gβ1, Gβ2 with the streptavidin binding domain plus an HA tag, and Gγ2 or Gγ7 with calmodulin binding domain plus a Flag tag allows purification of complexes specifically associated with unique Gβγ combinations.

9. TAP-tagged vector pIRESpuro-GLUE-N1 [1, 9].

10. Human embryonic kidney 293 cells (HEK 293T).

2.2 Isolation of Nuclei

1. PBS: Prepare a 10× stock of PBS: 1.37 M NaCl, 27 mM KCl, 100 mM Na_2HPO_4, 18 mM KH_2PO_4. Place a 1 L graduated cylinder or glass beaker on a stir plate; place a stir bar inside with approximately 500 mL of water. Weigh out 80 g NaCl, 2 g KCl, 14.4 g Na_2PO_4, 2.4 g KH_2PO_4. Cover beaker or cylinder with Parafilm to prevent contamination while mixing. Mix until dissolved. Add distilled water up to 1 L. Store at room temperature.

2. EDTA: Prepare 0.5 M EDTA pH 8.0. Add 18.61 g of disodium EDTA $2H_2O$ to 80 mL of water in a beaker or graduated cylinder on a stir plate. While stirring, add ~2 g of NaOH pellets (*see* **Note 2**). Adjust to 100 mL with H_2O. Store at room temperature.

3. EGTA: Prepare 0.2 M EGTA pH 7.5. Add 7.61 g of EGTA to 80 mL of sterile water on a stir plate. Adjust pH to 7.5 using NaOH pellets. Adjust to 100 mL with H_2O. Store at room temperature.

4. PBS plus EGTA and EDTA. For 200 mL, add 20 mL of 10× PBS to 170 mL of sterile water. Add 800 μL of 0.5 M EDTA pH 8.0 for a final concentration of 2 mM. Add 2 mL of EGTA pH 7.5 for a final concentration of 2 mM. Adjust volume to 200 mL. This solution should be made fresh, but can be stored at 4 °C for a few days.

5. 2 M $MgCl_2$: For 50 mL, add 20.33 g to 20 mL water. Mix until dissolved and then adjust volume to 50 mL. Store at room temperature.

6. 1 M HEPES-NaOH pH 8.0: For 50 mL, add 11.92 g of HEPES to 30 mL of water. Mix until dissolved. Check pH and add NaOH pellets until pH 8.0 is reached. Adjust volume to 50 mL with water. Store at room temperature.

7. 1 M DTT: For 10 mL, add 1.5 g of DTT to 8 mL of water. Adjust to 10 mL. Aliquot and freeze at –20 °C.

8. 100 mM PMSF: For 10 mL, add 174 mg of PMSF to 9 mL of isopropanol. Adjust to 10 mL. Aliquot and freeze at –20 °C (*see* **Note 3**).

9. 200 mM Na_3VO_4: For 10 mL add 0.368 g to 9 mL of water. Check pH and adjust to 10.0 using 1 M NaOH or 1 M HCl. Boil for 10 min or until solution clears. Cool on ice and adjust pH again. Boil the solution again and recheck pH after cooling. Repeat, adjusting pH as necessary, until pH stabilizes at 10.0. Aliquot and store at –20 °C.

10. Lysis Buffer: 320 mM sucrose, 10 mM HEPES-NaOH, 5 mM $MgCl_2$, 1 mM DTT, 1 mM PMSF, 1 mM Na_3VO_4, 1 % Triton X-100. For a 70 mL solution, weigh out 7.667 g of sucrose and add to 30 mL of water. Mix using a stir bar. Add 700 μL 1 M HEPES-NaOH, 175 μL 2 M $MgCl_2$, 70 μL 1 M DTT, 175 μL PMSF, 700 μL Na_3VO_4, 700 μL 100 % Triton X-100. Adjust volume to 70 mL. Prepare fresh, just prior to use (*see* **Note 4**).

11. High-sucrose buffer: 1.8 M sucrose, 10 mM HEPES-NaOH, 5 mM $MgCl_2$, 1 mM DTT, 1 mM PMSF, 1 mM Na_3VO_4, 1 % Triton X-100. Weigh out 43.127 g of sucrose. Prepare as stated above, heat solution to dissolve sucrose. Prepare fresh, just prior to use.

12. Resuspension buffer: 320 mM sucrose, 10 mM HEPES-NaOH, 5 mM $MgCl_2$, 1 mM DTT, 1 mM PMSF, 1 mM Na_3VO_4, 1 % Triton X-100, 1× Protease Inhibitor. For a 20 mL solution, weigh out 1.919 g of sucrose and add to 10 mL of water. Prepare as stated above. Prepare fresh, just prior to use.

2.3 Nuclear Lysis

1. 1 M HEPES-KOH pH 8.0: For 50 mL, add 11.92 g of HEPES to 30 mL of water. Mix until dissolved. Check pH and add KOH pellets until pH 8.0 is reached. Adjust volume to 50 mL with water. Store at room temperature.

2. 1 M KCl: For 50 mL, add 3.73 g of KCl to 40 mL water. Mix until dissolved. Adjust volume to 50 mL. Store at room temperature.

3. TAP lysis buffer: 1 mM $MgCl_2$, 10 % glycerol, 50 mM HEPES-KOH pH 8.0, 100 mM KCl, 0.1 % NP-40, 0.2 mM EDTA, 2 mM DTT, 1× Protease Inhibitor. For 50 mL combine 25 μL 2 M $MgCl_2$, 5 mL glycerol, 2.5 mL 1 M HEPES-KOH, 5 mL 1 M KCl, 71.43 μL 70 % NP-40, 50 μL 0.2 M EDTA, 100 μL 1 M DTT, 50 μL 1,000× Protease Inhibitor to 10 mL of water. Adjust to 50 mL. Prepare fresh.

4. Dounce tissue homogenizer, 2 mL capacity.

5. Benzonase® Nuclease (Sigma-Aldrich, Saint Louis, MO, USA).

2.4 Tandem Affinity Purification

1. Streptavidin Sepharose™ High Performance beads (GE Healthcare).

2. Calmodulin Sepharose™ 4B beads (GE Healthcare).

3. 5 M NaCl: For 50 mL, add 14.61 g of NaCl to 25 mL of water. Mix until dissolved. Adjust volume to 50 mL. Store at room temperature.

4. AcTEV™ (Tobacco Etch Virus) Protease (Invitrogen Life Technologies, Carlsbad, CA, USA).

5. TEV buffer: 150 mM NaCl, 10 mM HEPES-KOH pH 8.0, 0.5 mM EDTA, 0.1 % NP-40, 1 mM DTT. For 50 mL combine 1.5 mL 5 M NaCl, 500 μL HEPES-KOH, 125 μL 0.2 M EDTA, 71.43 μL 70 % NP-40, 50 μL 1 M DTT to 10 mL water. Adjust to 50 mL. Prepare fresh.

6. 1 M MgOAc: For 25 mL add 5.381 g of MgOAc to 15 mL of water. Adjust to 25 mL. Store at room temperature.

7. 1 M imidazole: For 10 mL add 0.6808 g of imidazole to 5 mL of water. Adjust to 10 mL. Can be stored at 4 °C protected from light for up 1 year.

8. 2 M $CaCl_2$: For 50 mL, add 7.351 g of $CaCl_2$ to 30 mL of water. Adjust to 50 mL. Store at room temperature.

9. Calmodulin binding buffer: 10 mM HEPES-KOH pH 8.0, 150 mM NaCl, 1 mM MgOAc, 1 mM imidazole, 0.1 % NP-40, 2 mM $CaCl_2$, 10 mM β-mercaptoethanol. For 50 mL add 500 μL 1 M HEPES-KOH, 1.5 mL 5 M NaCl, 50 μL 1 M MgOAc, 50 μL 1 M imidazole, 71.43 μL 70 % NP-40, 100 μL 2 M $CaCl_2$, 39 μL 14 M β-mercaptoethanol to 30 mL water. Adjust to 50 mL. Prepare fresh.

10. 1 M ammonium bicarbonate pH 8.0: For 25 mL add 98.8 mg of ammonium bicarbonate to 20 mL of water. Adjust to 25 mL with water (*see* **Note 5**). Prepare fresh.

11. Calmodulin rinsing buffer: 50 mM ammonium bicarbonate, 75 mM NaCl, 1 mM MgOAc, 1 mM imidazole, 2 mM $CaCl_2$. For 25 mL add 1,250 μL 1 M ammonium bicarbonate, 375 μL 5 M NaCl, 25 μL 1 M MgOAc, 25 μL 1 M imidazole, 50 μL 2 M $CaCl_2$ to 15 mL water. Adjust to 25 mL. Prepare fresh.

12. 0.5 M ammonium hydroxide: For 25 mL, add 0.86 mL of 14.534 M stock solution to 6.25 mL of water. Adjust to 25 mL with water. Prepare fresh.

13. Monoclonal antibody, mouse anti-HA.11, clone 16B12 (Covance, Montreal, QC, CA).

14. Nucleoporin p62 antibody, mouse, clone 53 (BD Transduction Laboratories™).

15. Gβ$_{1-4}$ antibody (T-20), rabbit (Santa Cruz Biotechnology, Inc., Santa Cruz, CA, USA).

3   Methods

All steps must be carried out on ice or at 4 °C unless otherwise indicated (*see* **Note 6**). Centrifugation steps must be performed at 4 °C. Samples (of approximately 50 μL) are taken at various stages of the protocol to ensure the bait protein binds and elutes from the beads. This is critical to determining whether the protocol has worked.

3.1   Tissue Culture

1. Grow 10 T175 flasks of HEK 293 cells to 40 % confluency in Dulbecco's MEM supplemented with FBS and P/S (*see* **Note 7**).

2. Transfect cells with TAP tag plasmid expressing Gβ$_1$ or the empty vector using Lipofectamine 2000 according to the manufacturer's instructions (*see* **Note 8**).

3. Harvest after 2 days as outlined below.

3.2   Nuclear Isolation

To ensure the isolation of intact nuclei, cells are lysed using a Triton X-100 based buffer and nuclei are pelleted through a 1.8 M sucrose cushion. This results in intact nuclei that can be viewed under the microscope. To confirm the purity of the nuclear fraction, we used antibodies against nucleoporin p62, a nucleoporin protein found exclusively in the nucleus, and against β-tubulin, found in the cytosol (Fig. 1) [24]. Endogenous Gβ was detected using anti-Gβ$_{1-4}$ antibody; tagged Gβ was detected with anti-HA antibody. To increase

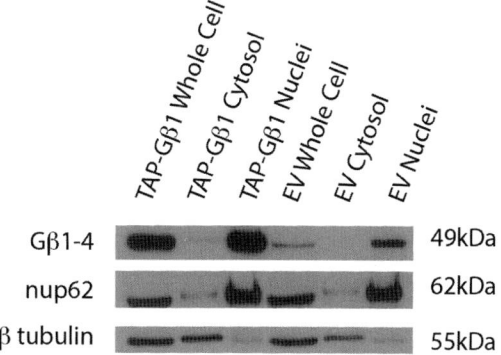

Fig. 1 Nuclear isolation of TAP-tagged Gβ$_1$ and empty vector (EV) transfected HEK 293 cells. Staining with Nup62 and β-tubulin antibodies was used to indicate the purity of the nuclear fraction. TAP-Gβ$_1$ is found in both the cytosolic and nuclear fractions

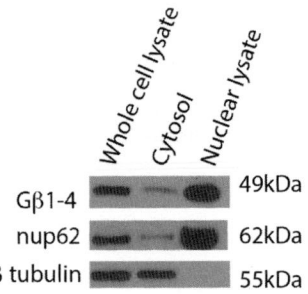

Fig. 2 Nuclear isolation of HEK 293 cells. Staining with Nup62 and β-tubulin antibodies was used to indicate the purity of the nuclear fraction. Endogenous Gβ is found in both the cytosolic and nuclear fractions

protein yield from nuclear isolations, they were done in 50 mL conical tubes. Due to increased protein expression in the cytosol compared with the nucleus, the cytosolic fraction was used as a positive control to confirm the success of purification, ensuring expression of the bait protein and proper binding and elution in all steps of the protocol. Endogenous Gβγ subunits are also localized to the nucleus (Fig. 2). The optimization of the protocol for use with nuclear lysates will more clearly reveal new interactions for Gβγ that occur specifically in the nucleus.

1. Remove flasks from incubator and place on ice (*see* **Note 9**).

2. Wash cells 2× with 10 mL PBS plus EGTA, EDTA (*see* **Note 10**).

3. Add another 10 mL of PBS plus EGTA, EDTA and tap the side of the flask to remove cells. Pipette up and down to remove cells from the surface. Remove cells from flasks and add to new 15 mL conical tubes on ice.

4. Centrifuge at $300 \times g$ for 5 min.

5. Discard supernatant.

6. Add 20 mL of lysis buffer to each tube and incubate on ice for 6 min (*see* **Note 11**).

7. While cells are incubating, add 15 mL of high sucrose buffer to new 50 mL conical tubes.

8. After 6 min, slowly place the lysed cells on top of the sucrose cushion being very careful not to disturb the barrier (*see* **Notes 12** and **13**).

9. Centrifuge at $4,600 \times g$ for 30 min at 4 °C (*see* **Note 14**).

10. Remove the top 18 mL cytosolic layer to a new 50 mL tube, being sure to collect only top layer, not the high sucrose cushion. Remove a sample for western blot. Proceed to tandem affinity purification step for cytosolic fraction or store at −80 °C (*see* **Notes 15** and **16**).

11. Slowly remove the rest of the cytosol followed by the high sucrose cushion. This can be discarded (*see* **Note 17**).

12. Resuspend the nuclear pellet in 1 mL of resuspension buffer. Place the 1 mL in a new clean Eppendorf tube. Centrifuge at $500 \times g$ for 5 min. Remove a sample for western blot.

13. Remove the resuspension buffer. If not immediately lysing the nuclei, resuspend in resuspension buffer and store at 4 °C. Remove a sample for western blot.

3.3 Nuclear Lysis

1. Resuspend all nuclei in 2 mL Tap lysis buffer with phosphatase and protease inhibitors (*see* **Note 18**).

2. Transfer to Dounce homogenizer and subject to ten passes. Transfer to new 2 mL tube (*see* **Note 19**).

3. Add 2 μL of benzonase (250 U/μL) and leave on a shaker overnight (*see* **Note 20**).

4. Centrifuge lysate at 17.6 g for 10 min to remove insoluble fraction (*see* **Note 21**).

3.4 Tandem Affinity Purification

The general workflow for TAP is shown in Fig. 3. In our hands, the standard TAP protocol works for whole cell lysates (Table 1), pulling down a number of known Gβγ-interacting proteins. Further, we noted Gβγ-interacting proteins identified from screens in whole cell lysates with either TAP-Gβ1 or the different split-TAP pairs include proteins with known functions in the nucleus (Table 2) further supporting a broader role for Gβγ in the nucleus. However, the standard Gβγ TAP protocol, as described by Ahmed et al., did not work with nuclear lysates [4]. Therefore, we have made changes that are outlined below. With this adjusted protocol (*see* Fig. 4), we were able to identify known interactors with $Gβ_1$ in both the cytosol and the nucleus (Tables 3 and 4).

1. Wash 2× 100 μL of streptavidin Sepharose beads 3× with 1 mL of TAP lysis buffer. Spin down beads at $200 \times g$ for 1 min.

2. Resuspend in 400 μL of TAP lysis buffer in 2 mL tube.

3. Combine the supernatant from nuclear lysis into a 15 mL tube and add washed beads. Add second set of beads to 50 mL of cytosolic fraction. Incubate on shaker overnight (*see* **Note 22**).

4. Spin down beads at $200 \times g$ for 1 min. Remove supernatant and save sample for western blot.

5. Wash beads (from cytosol and nuclei) 3× with TAP lysis buffer.

6. Wash beads 2× with TEV buffer. Save a sample of TEV wash for western blot.

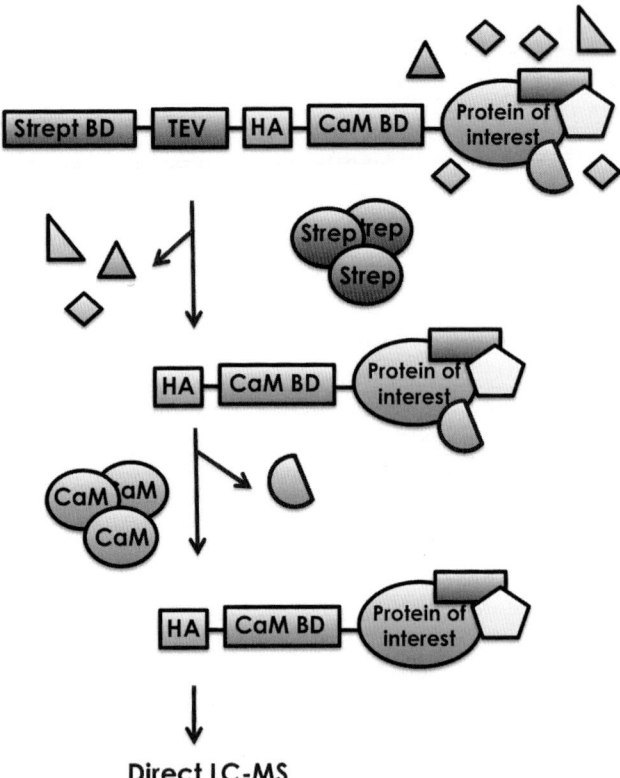

Fig. 3 Schema depicting the tandem affinity purification protocol. A calmodulin binding domain and streptavidin binding domain are fused to the protein of interest together or in the case of split-TAP alone. These tags allow binding to streptavidin and calmodulin beads. The tobacco etch virus (TEV) site allows for release of the bait protein from the streptavidin beads. The HA tag allows for identification of the fusion protein. In some split TAP constructs, Flag tags were used instead

7. Add 250 μL of TEV buffer to beads. Add 10 U of TEV and incubate overnight on shaker.

8. Spin down beads and remove supernatant to 2 mL tube. Add 300 μL of calmodulin binding buffer to streptavidin beads. Mix by inverting tube. Spin down and collect supernatant. Repeat 2×, adding each wash to the streptavidin elution. Save a sample for western blot.

9. Add 1/250 volume (5 μL) of 2 M $CaCl_2$ to tube for a final concentration of 1 mM.

10. Spin down to remove residual streptavidin beads that may have transferred over.

11. Remove supernatant to new tube.

12. Wash calmodulin Sepharose beads 3× with calmodulin binding buffer.

Table 1
Proteins pulled down from Gβ₁ from whole cell lysates

Bait protein (TAP or split-TAP)	Protein ID	Description	Unique peptides
Gβ1γ2, Gβ1γ7, Gβ2γ7, Gβ2γ2	GNA11	Guanine nucleotide binding protein (G protein), alpha 11 (Gq class)	11, 2, 5, 4
Gβ1γ2, Gβ1γ7, Gβ2γ7, Gβ2γ2	GNA13	Guanine nucleotide binding protein (G protein), alpha 13	11, 4, 5, 7
Gβ1γ2, Gβ2γ7, Gβ2γ2	GNAI2	Guanine nucleotide binding protein (G protein), alpha inhibiting activity polypeptide 2	3, 1, 2
Gβ1γ2, Gβ1γ7, Gβ2γ7, Gβ2γ2	GNAI3	Guanine nucleotide binding protein (G protein), alpha inhibiting activity polypeptide 3	7, 3, 6, 3
Gβ1γ2, Gβ2γ2	GNAI1	Guanine nucleotide binding protein (G protein), alpha inhibiting activity polypeptide 1	3, 1
Gβ1γ2, Gβ1γ7, Gβ2γ7, Gβ2γ2	GNAQ	Guanine nucleotide binding protein (G protein), q polypeptide	5, 2, 3, 6
Gβ1γ2	GNAS	Guanine nucleotide binding protein (G protein), alpha stimulating activity polypeptide 1	5
Gβ1γ2	GNAZ	Guanine nucleotide binding protein (G protein), alpha z polypeptide	3
Gβ2γ2	GNAO1	Guanine nucleotide binding protein (G protein), alpha activating activity polypeptide O	1
Gβ1γ2, Gβ1γ7, Gβ2γ7	GNB1	Guanine nucleotide binding protein (G protein), beta polypeptide 1	25, 18, 8
Gβ1γ2	GNB2	Guanine nucleotide binding protein (G protein), beta polypeptide 2	3
Gβ1γ7	GNB3	Guanine nucleotide binding protein (G protein), beta polypeptide 3	1
Gβ1γ2, Gβ2γ7	GNG12	Guanine nucleotide binding protein (G protein), gamma 12	4, 1
Gβ1γ2, Gβ2γ2	GNG2	Guanine nucleotide binding protein (G protein), gamma 2	7, 2
Gβ1γ2	GNG5	Guanine nucleotide binding protein (G protein), gamma 5	3
Gβ1γ2, Gβ1γ7, Gβ2γ7, Gβ2γ2	GNG7	Guanine nucleotide binding protein (G protein), gamma 7	1, 3, 2, 1
Gβ1γ2, Gβ1γ7, Gβ2γ7, Gβ2γ2	CCT2	Chaperonin containing TCP1, subunit 2 (beta)	11, 8, 4, 6
Gβ1γ2, Gβ1γ7, Gβ2γ7, Gβ2γ2	CCT3	Chaperonin containing TCP1, subunit 3 (gamma)	9, 9, 4, 4
Gβ1γ2, Gβ1γ7, Gβ2γ2	CCT4	Chaperonin containing TCP1, subunit 4 (delta)	10, 5, 2

(continued)

Table 1
(continued)

Bait protein (TAP or split-TAP)	Protein ID	Description	Unique peptides
Gβ1γ2, Gβ1γ7, Gβ2γ7, Gβ2γ2	CCT5	Chaperonin containing TCP1, subunit 5 (epsilon)	6, 6, 2, 4
Gβ1γ2, Gβ1γ7, Gβ2γ7, Gβ2γ2	CCT6A	Chaperonin containing TCP1, subunit 6A (zeta 1)	8, 9, 2, 4
Gβ1γ2, Gβ1γ7, Gβ2γ2	CCT7	Chaperonin containing TCP1, subunit 7 (eta)	10, 4, 2
Gβ1γ2, Gβ1γ7, Gβ2γ7, Gβ2γ2	CCT8	Chaperonin containing TCP1, subunit 8 (theta)	14, 7, 5, 6
Gβ1γ2, Gβ1γ7, Gβ2γ2	TCP1	T-complex 1	10, 6, 2
Gβ1γ2	KCTD2	Potassium channel tetramerization domain containing 2	9
Gβ1γ2	KCTD12	Potassium channel tetramerization domain containing 12	11
Gβ1γ2	KCTD16	Potassium channel tetramerization domain containing 16	1
Gβ1γ2	KCTD17	Potassium channel tetramerization domain containing 17	5
Gβ1γ2, Gβ1γ7, Gβ2γ7, Gβ2γ2	KCTD5	Potassium channel tetramerization domain containing 5	14, 2, 3, 3
Gβ1γ2, Gβ1γ7, Gβ2γ7, Gβ2γ2	RADIL	Ras association and DIL domains	24, 5, 10, 6
Gβ1γ2, Gβ1γ7, Gβ2γ7, Gβ2γ2	PDCL	Phosducin-like	4, 1, 7, 5
Gβ1γ2	RAP1A	Member of RAS oncogene family	4
Gβ1γ2, Gβ2γ7, Gβ2γ2	RAP1B	Member of RAS oncogene family	1, 1, 1

Split TAP combinations are listed as $G\beta_x\gamma_y$, single TAP constructs as $G\beta_x$ or $G\gamma_y$

13. Resuspend beads in 400 μL of calmodulin binding buffer and transfer to streptavidin elution.

14. Incubate overnight on shaker.

15. Spin down beads and remove supernatant. Save this sample for western blot.

16. Wash beads 3× with 1 mL calmodulin binding buffer. Save a sample of the wash for western blot.

17. Wash beads 3× with calmodulin rinsing buffer. Save a sample for western blot.

18. Add 150 μL 0.5 M ammonium hydroxide. Leave on bench for 5 min.

Table 2
Nuclear proteins pulled down from various Gβ and Gγ combinations

Bait protein (TAP or split-TAP)	Protein ID	Description	Unique peptides
Gβ1γ2	IPO7	Importin 7	2
Gβ1γ2, Gβ1, Gγ2	KPNB1	Karyopherin β1	1, 5, 1
Gγ2	EXP1	Exportin 1	1
Gγ2	EXP5	Exportin 5	1
Gβ2γ7	NPM1	Nucleophosmin 1	2
Gβ1γ2, Gγ2, Gβ2γ2, Gβ2γ7	PRKDC	Protein kinase, DNA activated, catalytic polypeptide isoform 1	8, 50, 2, 6
Gβ1γ2	NCOA5	Nuclear receptor coactivator 5	1
Gβ2γ7, Gβ2γ2	NFKB	NF-κB	1, 1
Gγ2	HNRNH1	Heterogeneous nuclear ribonucleoprotein H1	1

Split TAP combinations are listed as Gβ$_x$γ$_y$, single TAP constructs as Gβ$_x$ or Gγ$_y$

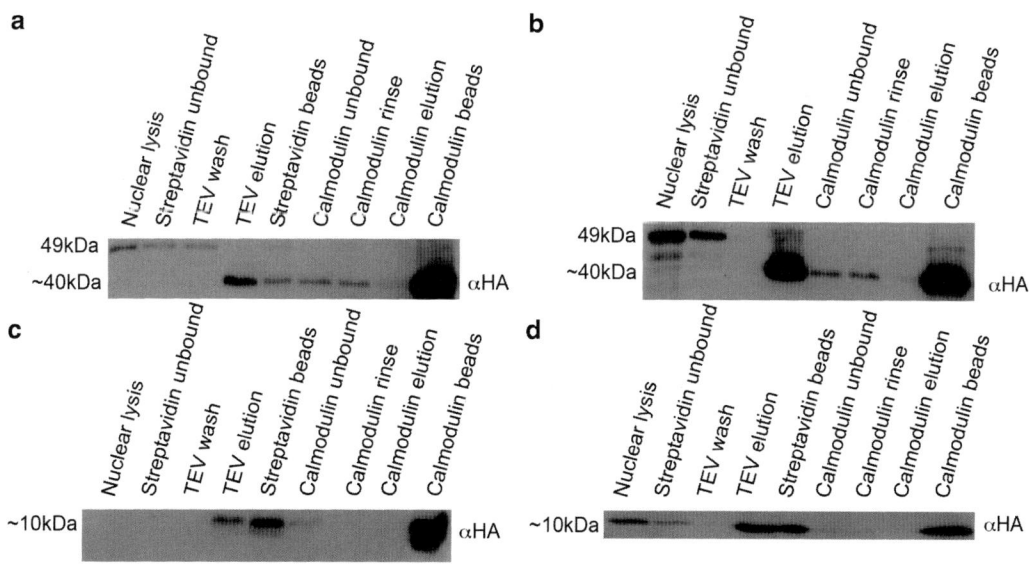

Fig. 4 TAP of Gβ$_1$ from nuclear and cytosolic fraction. Panel (**a**) is from the cytosolic fraction of HEK 293 cells transfected with the TAP-Gβ$_1$. Panel (**b**) is from the nuclear fraction of HEK 293 cells transfected with TAP-Gβ$_1$. Panel (**c**) represents the cytosolic fraction of HEK 293 cells transfected with empty vector. Panel (**d**) is from the nuclear fraction of HEK 293 cells transfected with empty vector. Gβ$_1$ is bound and eluted from the streptavidin and calmodulin beads and runs at 49 kDa. The HA tag from empty vector transfected cells runs at ~10 kDa and can be seen on the gel

Table 3
Proteins pulled down from Gβ_1 following nuclear isolation

Bait protein	Protein ID	Description	Unique peptides
Gβ1γ7	HNRNPM	Heterogeneous nuclear ribonucleoprotein M	27
Gβ1γ7, Gβ1	GNB1	Guanine nucleotide binding protein (G Protein), beta polypeptide 1	26, 12
Gβ1γ7	GNB2	Guanine nucleotide binding protein (G protein), beta polypeptide 2	25
Gβ1γ7, Gβ1	GNA13	Guanine nucleotide binding protein (G protein), alpha 13	7, 2
Gβ1γ7, Gβ1	GNAS	Guanine nucleotide binding protein (G Protein), alpha stimulating activity polypeptide 1	6, 2
Gβ1γ7	GNA11	Guanine nucleotide binding protein (G protein), alpha 11 (Gq class)	5
Gβ1γ7	CKAP4	Cytoskeleton-associated protein 4	5
Gβ1γ7	RADIL	Ras association and DIL domains	4
Gβ1γ7	PHB2	Prohibitin 2	4
Gβ1γ7	GNAI3	Guanine nucleotide binding protein (G protein), alpha inhibiting activity polypeptide 3	4
Gβ1	HNRNPM	Heterogeneous nuclear ribonucleoprotein M	4
Gβ1	DNAJC25-GNG10	DNAJC25-GNG10 readthrough transcript	4
Gβ1	ACTBL2	Actin, beta-like 2	4
Gβ1γ7, Gβ1	GNG7	Guanine nucleotide binding protein (G protein), gamma 7	3, 1
Gβ1	FBLN5	Fibulin 5	3
Gβ1γ7	KCTD12	Potassium channel tetramerization domain containing 12	3
Gβ1γ7	HNRNPF	Heterogeneous nuclear ribonucleoprotein F	3
Gβ1γ7	MATR3	Matrin 3	3
Gβ1γ7	HNRNPU	Heterogeneous nuclear ribonucleoprotein U (scaffold attachment factor A)	2
Gβ1γ7	MYBBP1A	MYB binding protein (P160) 1a	2
Gβ1γ7	RBM12	RNA binding motif protein 12	2
Gβ1γ7	ATAD3B	ATPase family, AAA domain containing 3B	2
Gβ1	HIST1H1C	Histone cluster 1, H1c	2
Gβ1	ANXA5	Annexin A5	2
Gβ1	AKR7A2	Aldo-keto reductase family 7, member A2 (aflatoxin aldehyde reductase)	2

(continued)

Table 3
(continued)

Bait protein	Protein ID	Description	Unique peptides
Gβ1	GNAI2	Guanine nucleotide binding protein (G protein), alpha inhibiting activity polypeptide 2	2
Gβ1	FAM82B	Regulator of microtubule dynamics 1	2
Gβ1γ7	RFC1	Replication factor C (activator 1) 1, 145 kDa	1
Gβ1γ7, Gβ1	HNRNPD	Heterogeneous nuclear ribonucleoprotein D (AU-rich element RNA binding protein 1, 37 kDa)	1, 1
Gβ1γ7	BSG	Basigin	1
Gβ1γ7	ANXA6	Annexin A6	1
Gβ1γ7	PTPLAD1	Protein tyrosine phosphatase-like A domain containing 1	1
Gβ1γ7	RFC2	Replication factor C (activator 1) 2, 40 kDa	1
Gβ1γ7	TESC	Tescalcin	1
Gβ1γ7	HIST1H2AI	Histone cluster 1, H2ai	1
Gβ1γ7	RPS27A	Ribosomal protein S27a	1
Gβ1γ7	HNRNPA3	Heterogeneous nuclear ribonucleoprotein A3	1
Gβ1γ7	PHB	Prohibitin	1
Gβ1γ7	KCTD5	Potassium channel tetramerization domain containing 5	1
Gβ1	GNG12	Guanine nucleotide binding protein (G protein), gamma 12	1
Gβ1	SPA17	Sperm autoantigenic protein 17	1
Gβ1	GNAQ	Guanine nucleotide binding protein (G protein), q polypeptide	1
Gβ1	MTERFD2	MTERF domain containing 2	1

Split TAP combinations are listed as $G\beta_x\gamma_y$, single TAP constructs as $G\beta_x$ or $G\gamma_y$

Table 4
Proteins pulled down from Gβ₁ following cytosolic isolation

Bait protein	Protein ID	Description	Unique peptides
Gβ1, Gβ1γ7	GNAI3	Guanine nucleotide binding protein (G protein), alpha inhibiting activity polypeptide 3	88, 33
Gβ1, Gβ1γ7	GNAI2	Guanine nucleotide binding protein (G protein), alpha inhibiting activity polypeptide 2	77, 23
Gβ1, Gβ1γ7	GNAI1	Guanine nucleotide binding protein (G protein), alpha inhibiting activity polypeptide 1	67, 24

(continued)

Table 4
(continued)

Bait protein	Protein ID	Description	Unique peptides
Gβ1	CAPNS1	Calpain, small subunit 1	35
Gβ1γ7, Gβ1, Gβ1	GNB1	Guanine nucleotide binding protein (G protein), beta polypeptide 1	34, 20, 4
Gβ1	MYLK2	Myosin light chain kinase 2	32
Gβ1, Gβ1γ7	GNAS	Guanine nucleotide binding protein (G protein), alpha stimulating activity polypeptide 1	30, 17
Gβ1γ7, Gβ1	GNA13	Guanine nucleotide binding protein (G protein), alpha 13	28, 28
Gβ1	GNAQ	Guanine nucleotide binding protein (G protein), q polypeptide	28
Gβ1, Gβ1γ7	GNA11	Guanine nucleotide binding protein (G protein), alpha 11 (Gq class)	28, 14
Gβ1γ7	DSP	Desmoplakin	26
Gβ1, Gβ1γ7, Gβ1	TUBB4B	Tubulin, beta 4B class IVb	26, 8, 4
Gβ1, Gβ1	PDCL	Phosducin-like	20, 5
Gβ1γ7	GNAO1	Guanine nucleotide binding protein (G protein), alpha activating activity polypeptide O	20
Gβ1γ7, Gβ1	KCTD2	Potassium channel tetramerization domain containing 2	19, 7
Gβ1γ7, Gβ1	KCTD12	Potassium channel tetramerization domain containing 12	17, 9
Gβ1γ7	GNAQ	Guanine nucleotide binding protein (G protein), q polypeptide	15
Gβ1γ7, Gβ1	KCTD5	Potassium channel tetramerization domain containing 5	15, 5
Gβ1γ7	HNRNPA2B1	Heterogeneous nuclear ribonucleoprotein A2/B1	13
Gβ1, Gβ1γ7	GNG7	Guanine nucleotide binding protein (G protein), gamma 7	10, 2
Gβ1γ7	TUBB	Tubulin, beta class I	9
Gβ1γ7	SERPINB3	Serpin peptidase inhibitor, clade B (ovalbumin), member 3	8
Gβ1γ7	JUP	Junction plakoglobin	8
Gβ1γ7	ALB	Albumin	8
Gβ1, Gβ1γ7	GNAZ	Guanine nucleotide binding protein (G protein), alpha z polypeptide	7, 2
Gβ1	CPNE1	Copine I	7
Gβ1	TP53I11	Tumor protein p53 inducible protein 11	7
Gβ1	MYL6	Myosin, light chain 6, alkali, smooth muscle and non-muscle	7
Gβ1	HPCAL1	Hippocalcin-like 1	6
Gβ1	GNG11	Guanine nucleotide binding protein (G protein), gamma 11	6

(continued)

**Table 4
(continued)**

Bait protein	Protein ID	Description	Unique peptides
Gβ1	IPO5	Importin 5	5
Gβ1	PDCD6	Programmed cell death 6	5
Gβ1	CCT3	Chaperonin containing TCP1, subunit 3 (gamma)	5
Gβ1γ7	DSG1	Desmoglein 1	5
Gβ1γ7	TGM3	Transglutaminase 3	5
Gβ1	ATAD3A	ATPase family, AAA domain containing 3A	4
Gβ1	PHB2	Prohibitin 2	4
Gβ1	CAND1	Cullin-associated and neddylation-dissociated 1	4
Gβ1	CPNE3	Copine III	4
Gβ1γ7	HSPA1L	Heat shock 70 kDa protein 1-like	4
Gβ1γ7	FUS	Fused in sarcoma	4
Gβ1γ7	AMOT	Angiomotin	4
Gβ1γ7	CAP14	Catabolite activator gene protein	4
Gβ1γ7, Gβ1	SLC25A5	Solute carrier family 25 (mitochondrial carrier; adenine nucleotide translocator), member 5	4, 2
Gβ1γ7	CCT5	Chaperonin containing TCP1, subunit 5 (epsilon)	4
Gβ1	PPP3R1	Protein phosphatase 3, regulatory subunit B, alpha	3
Gβ1	PTPLAD1	Protein tyrosine phosphatase-like A domain containing 1	3
Gβ1	CKAP4	Cytoskeleton-associated protein 4	3
Gβ1	ERLIN2	ER lipid raft associated 2	3
Gβ1	IQGAP1	IQ motif containing GTPase activating protein 1	3
Gβ1	CHP1	Calcineurin-like EF-hand protein 1	3
Gβ1	ST13	Suppression of tumorigenicity 13 (colon carcinoma) (Hsp70 interacting protein)	3
Gβ1	MAPK1	Mitogen-activated protein kinase 1	3
Gβ1	VPS35	Vacuolar protein sorting 35 homolog (*S. cerevisiae*)	3
Gβ1 Gβ1γ7	CCT4	Chaperonin containing TCP1, subunit 4 (delta)	3, 2
Gβ1	ANXA1	Annexin A1	3
Gβ1	ACTBL2	Actin, beta-like 2	3
Gβ1	MYL12A	Myosin, light chain 12A, regulatory, non-sarcomeric	3
Gβ1	FBLN5	Fibulin 5	3
Gβ1	CCT5	Chaperonin containing TCP1, subunit 5 (epsilon)	3

(continued)

Table 4
(continued)

Bait protein	Protein ID	Description	Unique peptides
Gβ1γ7	HNRNPA1	Heterogeneous nuclear ribonucleoprotein A1	3
Gβ1γ7	LGALS7	Lectin, galactoside-binding, soluble, 7	3
Gβ1γ7	GAPDH	Glyceraldehyde-3-phosphate dehydrogenase	3
Gβ1γ7	ANAX2	Transient receptor potential cation channel, subfamily V, member 5	3
Gβ1γ7	SERPINB12	Serpin peptidase inhibitor, clade B (ovalbumin), member 12	3
Gβ1γ7	IDE	Insulin-degrading enzyme	3
Gβ1	ATP6V1D	ATPase, H + transporting, lysosomal 34 kDa, V1 subunit D	2
Gβ1, Gβ1γ7	PHB	Prohibitin	2, 1
Gβ1	HADHA	Hydroxyacyl-CoA dehydrogenase/3-ketoacyl-CoA thiolase/ enoyl-CoA hydratase, alpha subunit	2
Gβ1	CAPN2	Calpain 2, (m/II) large subunit	2
Gβ1	TLN1	Talin 1	2
Gβ1	CUL3	Cullin 3	2
Gβ1	MYO6	Myosin VI	2
Gβ1	TMEM109	Transmembrane protein 109	2
Gβ1	YWHAZ	Tyrosine 3-monooxygenase/tryptophan 5-monooxygenase activation protein, zeta polypeptide	2
Gβ1	S100A4	S100 calcium binding protein A4	2
Gβ1	MYO1B	Myosin IB	2
Gβ1	SPTAN1	Spectrin, alpha, non-erythrocytic 1	2
Gβ1, Gβ1γ7	CCT6A	Chaperonin containing TCP1, subunit 6A (zeta 1)	2, 2
Gβ1, Gβ1, Gβ1γ7	GNG12	Guanine nucleotide binding protein (G protein), gamma 12	3, 2, 2
Gβ1γ7	HSPB1	Heat shock 27 kDa protein 1	2
Gβ1γ7	DSC1	Desmocollin 1	2
Gβ1γ7	KCTD17	Potassium channel tetramerization domain containing 17	2
Gβ1γ7	RPS27A	Ribosomal protein S27a	2
Gβ1γ7	TXN	Thioredoxin	2
Gβ1γ7	PLEC	Plectin	2
Gβ1	DOLPP1	Dolichyldiphosphatase 1	1
Gβ1	COX4I1	Cytochrome c oxidase subunit IV isoform 1	1

(continued)

Table 4
(continued)

Bait protein	Protein ID	Description	Unique peptides
Gβ1	LMNB1	Lamin B1	1
Gβ1	SNX1	Sorting nexin 1	1
Gβ1	GNG4	Guanine nucleotide binding protein (G protein), gamma 4	1
Gβ1	VPS26A	Vacuolar protein sorting 26 homolog A (S. pombe)	1
Gβ1	HADHB	Hydroxyacyl-CoA dehydrogenase/3-ketoacyl-CoA thiolase/enoyl-CoA hydratase (trifunctional protein), beta subunit	1
Gβ1	G6PD	Glucose-6-phosphate dehydrogenase	1
Gβ1	TCP1	T-complex 1	1
Gβ1	RPLP2	Ribosomal protein, large, P2	1
Gβ1	CSTB	Cystatin B (stefin B)	1
Gβ1γ7	SBSN	Suprabasin	1
Gβ1γ7	COX7A2	Cytochrome c oxidase subunit VIIa polypeptide 2 (liver)	1
Gβ1γ7	CTSA	Cathepsin A	1
Gβ1γ7	HNRNPAB	Heterogeneous nuclear ribonucleoprotein A/B	1
Gβ1γ7	HNRNPA3	Heterogeneous nuclear ribonucleoprotein A3	1
Gβ1γ7	CCT7	Chaperonin containing TCP1, subunit 7 (eta)	1
Gβ1γ7	PABPC1	Poly(A) binding protein, cytoplasmic 1	1
Gβ1γ7	CDC42BPA	CDC42 binding protein kinase alpha (DMPK-like)	1
Gβ1γ7	RPS20	Ribosomal protein S20	1
Gβ1γ7	CDSN	Corneodesmosin	1

Split TAP combinations are listed as $G\beta_x\gamma_y$, single TAP constructs as $G\beta_x$ or $G\gamma_y$

19. Spin down beads and remove supernatant to new tube. Save 30 µL for silver stain gel and a sample for western blot.

20. Repeat 2× and combine supernatant.

21. Lyophilize supernatant for approximately 2 h (*see* **Note 23**).

22. Add 600 µL of water to the pellet to wash.

23. Lyophilize for another 2 h.

24. Send lyophilized sample for MS, run samples on a polyacrylamide gel followed by silver stain and western blot to detect bait protein.

3.5 Silver Stain
Silver stain allows detection of protein at concentrations in the nanogram range. If protein can be detected on a silver-stained gel, there should be sufficient protein eluted from the beads to be sent for MS. The protocol of choice should be determined based on whether the gel is used for identifying protein, or whether protein will be cut out of the gel and sent for MS [27].

1. Load 30 µL from first elution.

2. Run gel at 120 V for 60 min.

3. Stain gel with silver stain [28] (*see* **Note 24**).

4. Send sample for LC/MS [1].

5. Tandem mass spectra were searched against a FASTA file containing the human NCBI sequences as previously described [9]. The resulting peptide identifications were compared to control data and data from the "Crapome" database [28]. Protein identifications found in less than "60 Crapome" control experiments were determined to be positive hits.

Protein identifications from cytosolic fractions indicate that $G\beta_1$ can bind to known $G\beta$ interactors such as the $G\alpha$ subunits and $G\gamma$ subunits in both the cytosolic and nuclear fractions (Tables 2, 3, and 4). Therefore our $G\beta_1$ construct is functional as it pulls down previously characterized $G\beta_1$ protein interactors. In both our whole cell and nuclear isolation screens, we have been able to identify importin and exportin proteins (Tables 1 and 3). This may suggest a possible mechanism for import of $G\beta\gamma$ into the nucleus. Interesting interactors with $G\beta_1$ found through our screens include heterogeneous nuclear ribonuclear proteins, RNA binding motif protein, and replication factors (Table 3). Although these interactors remain to be validated, this data, along with accumulated evidence of $G\beta\gamma$ subunit interactions with transcription factors, point to a broad role for $G\beta\gamma$ in regulating transcription in the nucleus [19, 23–25]. Further research in our lab will focus on replicating and validating these results, allowing characterization of $G\beta\gamma$ signaling networks important in gene transcription.

4 Notes

1. Puromycin and hygromycin are used to select for TAP plasmids expressing tagged bait proteins in stable cell lines. Other appropriate antibiotics to select for the TAP plasmid can be used.

2. EDTA/EGTA will not dissolve until pH of 8.0 is reached. There is a small window of pH in which EDTA/EGTA will dissolve; therefore, add NaOH pellet very slowly while stirring and check pH frequently.

3. PMSF is rapidly inactivated in water. Therefore it should always be added to solutions just before use.

4. Depending on the number of T175 flasks, you will need to calculate the amount of lysis buffer, high sucrose buffer, and resuspension buffer required.

5. Ammonium bicarbonate must be made fresh for each experiment as the pH changes rapidly over time. Check pH using pH strips to avoid contamination. Do not adjust the pH when making ammonium bicarbonate; the solution should be at pH 8.0 when made if the concentration is correct.

6. If storing samples at any point, flash freeze them in liquid nitrogen and store at −80 °C. This will ensure that protein stability is maintained.

7. The binding capacity of streptavidin Sepharose beads (GE Healthcare) is 0.6 mg of biotinylated BSA/100 μL of medium. However, due to the number of wash steps involved in this protocol, we found that the optimal starting amount of protein is 100 mg. This translates into 10 or 20 T175 flasks of HEK 293 cells, respectively, for cytosolic or nuclear preparations.

8. Our lab has found that transfecting at much lower concentrations than recommended by the manufacturer is appropriate for HEK 293 cells. This should be optimized based on the cell line used and whether you will be using stable or transiently transfected cell lines. In our lab, we have generated stable cell lines for use with the TAP protocol to ensure levels of protein expression closer to that of the endogenous protein.

9. Cell detachment and lysis must be done on ice, large dishes or containers can be filled with ice, and tissue culture flasks placed on top during the process.

10. EGTA and EDTA chelate calcium and magnesium, which are required by integrins for surface adhesion [29]. Therefore, the use of EGTA and EDTA facilitates cell detachment. If cells are hard to detach, trypsin can be used in combination with EDTA.

11. The 6 min incubation time on ice is very important; if cells are lysed for longer, the amount of intact nuclei isolated will be decreased.

12. Slowly pipette the lysed solution down the side of the conical tube. This will prevent the two layers from mixing.

13. The 1.8 M sucrose cushion allows for enrichment of nuclei from HEK 293 cells, by preventing the cytosol and cell membrane from passing through the cushion during centrifugation. Intact nuclei pass through the cushion and are found as a pellet at the bottom of the tube.

14. This centrifuge speed is optimal to pellet the nuclei of HEK 293 cells. If using another cell type, this may differ.

15. There will be more protein isolated from the cytosol than the nuclei. Because the cytosol is diluted compared with the nuclei, combine the cytosol in a 50 mL conical tube and add the streptavidin beads to that.

16. Protein that will be used for further steps of the TAP protocol must be stored at −80 °C to ensure it will bind to the beads when thawed.

17. It is very important to remove the cytosolic fraction before removing the high sucrose cushion. This ensures a clean nuclear pellet with low levels of cytosolic contamination.

18. Phosphatase and protease inhibitors that are appropriate for your cells can be added. If using an inhibitor cocktail that contains 4-(2-aminoethyl) benzenesulfonyl fluoride hydrochloride (AEBSF), PMSF can be omitted as they have similar specificities.

19. To ensure that proper lysis of nuclei has occurred, place 10 µL of nuclear lysis on a glass slide and look at under light microscope with 10× magnification. You should not see any intact nuclei.

20. This amount will need to be optimized depending on the cell type used. You can test out differing concentrations of benzonase by determining the amount of DNA left (measure absorbance at 260 nm). It is very important to digest the DNA in the sample as it will be very highly concentrated and will affect the ability of nuclear proteins to bind to beads. Take a sample for western blot analysis following digestion with benzonase. Benzonase is effective at temperatures from 0 to 42 °C, Mg^{2+} concentrations of 1 to 10 mM, at pH 6.0–10.0, 0–100 mM DTT, and 0–150 mM monovalent cation (Na^+, K^+) concentrations. Because the buffer and temperatures used in this protocol are not optimal for benzonase activity, lysis should be left overnight using higher concentrations of benzonase as per manufacturer's instructions.

21. This centrifugation step allows separation of DNA bound proteins from soluble nuclear proteins. If this step is removed, the beads will clump together with the non-soluble fraction, preventing proper elution of protein from the beads in the following step.

22. The cytosolic sample does not require DNA digestion with benzonase. While nuclear sample is incubating with benzonase overnight, the cytosolic sample can remain incubating with streptavidin beads on the shaker or flash frozen in liquid nitrogen and stored at −80 °C for later use.

23. For further details on preparing proteins for MS analysis *see* Ahmed et al. [9].

24. If a band is seen following silver stain, there will be enough protein in the sample for detection by mass spectrometry. Otherwise, the protocol should be done again with more starting material.

References

1. Roy SJ, Glazkova I, Fréchette L et al (2013) Novel, gel-free proteomics approach identifies RNF5 and JAMP as modulators of GPCR stability. Mol Endocrinol 27:1245–1266

2. Rigaut G, Shevchenko A, Rutz B et al (1999) A generic protein purification method for protein complex characterization and proteome exploration. Nat Biotechnol 17:1030–1032

3. Puig O, Caspary F, Rigaut G et al (2001) The tandem affinity purification (TAP) method: a general procedure of protein complex purification. Methods 24:218–229

4. Ahmed SM, Daulat AM, Angers S (2011) Tandem affinity purification and identification of heterotrimeric G protein-associated proteins. Methods Mol Biol 756:357–370

5. Gloeckner CJ, Boldt K, Ueffing M (2009) Strep/FLAG tandem affinity purification (SF-TAP) to study protein interactions. In: Coligan JE (ed) Current protocols in protein science, Unit19.20, Chapter 19

6. Stamsas GA, Havarstein LS, Straume D (2013) CHiC, a new tandem affinity tag for the protein purification toolbox. J Microbiol Methods 92:59–63

7. Okada Y, Takano TY, Kobayashi NY et al (2011) New protein purification system using gold-magnetic beads and a novel peptide tag, "the methionine tag". Bioconjug Chem 22:887–893

8. Maine GN, Gluck N, Zaidi IW et al. (2009) Bimolecular affinity purification (BAP): tandem affinity purification using two protein baits. Cold Spring Harb Prot. pdb.prot5318

9. Ahmed SM, Daulat AM, Meunier A et al (2010) G protein βγ subunits regulate cell adhesion through Rap1a and its effector Radil. J Biol Chem 285:6538–6551

10. Chong S, Mersha FB, Comb DG et al (1997) Single-column purification of free recombinant proteins using a self-cleavable affinity tag derived from a protein splicing element. Gene 192:271–281

11. Kim JS, Raines RT (1993) Ribonuclease S-peptide as a carrier in fusion proteins. Protein Sci 2:348–356

12. Korndörfer IP, Skerra A (2002) Improved affinity of engineered streptavidin for the strep-tag II peptide is due to a fixed open conformation of the lid-like loop at the binding site. Protein Sci 11:883–893

13. Li Y (2010) Commonly used tag combinations for tandem affinity purification. Biotechnol Appl Biochem 55:73–83

14. Raines RT, McCormick M, Van Oosbree TR et al (2000) The S-tag fusion system for protein purification. Methods Enzymol 326:362–376

15. Stirling DA, Petrie A, Pulford DJ et al (1992) Protein A-calmodulin fusions: a novel approach for investigating calmodulin function in yeast. Mol Microbiol 6:703–713

16. Stofko-Hahn RE, Carr DW, Scott JD (1992) A single step purification for recombinant proteins. Characterization of a microtubule associated protein (MAP 2) fragment which associates with the type II cAMP-dependent protein kinase. FEBS Lett 302:274–278

17. Uhlén M, Guss B, Nilsson B et al (1984) Lindberg, expression of the gene encoding protein A in Staphylococcus aureus and coagulase-negative staphylococci. J Bacteriol 159:713–719

18. Lavallée-Adam M, Rousseau J, Domecq C et al (2013) Discovery of cell compartment specific protein-protein interactions using affinity purification combined with tandem mass spectrometry. J Proteome Res 12:272–281

19. Khan SM, Sleno R, Gora S et al (2013) The expanding roles of Gβγ subunits in G protein-coupled receptor signaling and drug action. Pharmacol Rev 65:545–577

20. Crouch MF (1991) Growth factor-induced cell division is paralleled by translocation of Giα to the nucleus. FASEB J 5:200–206

21. Tadevosyan A, Vaniotis G, Allen BG et al (2012) G protein-coupled receptor signalling in the cardiac nuclear membrane: evidence and possible roles in physiological and pathophysiological function. J Physiol 590:1313–1330

22. Vaniotis G, Allen BG, Hébert TE (2011) Nuclear GPCRs in cardiomyocytes: an insider's view of β-adrenergic receptor signaling. Am J Physiol Heart Circ Physiol 301:H1754–H1764

23. Spiegelberg BD, Hamm HE (2005) Gβγ binds histone deacetylase 5 (HDAC5) and inhibits its transcriptional co-repression activity. J Biol Chem 280:41769–41776

24. Robitaille M, Gora S, Wang Y et al (2010) Gβγ is a negative regulator of AP-1 mediated transcription. Cell Signal 22:1254–1266

25. Park JG, Muise A, He GP et al (1999) Transcriptional regulation by the γ5 subunit of a heterotrimeric G protein during adipogenesis. EMBO J 18:4004–4012

26. Keller BO, Sui J, Young AB et al (2008) Interferences and contaminants encountered in modern mass spectrometry. Anal Chim Acta 627:71–81

27. Chevallet M, Luche S, Rabilloud T (2006) Silver staining of proteins in polyacrylamide gels. Nat Protoc 1:1852–1858

28. Mellacheruvu D, Wright Z, Couzens AL et al (2013) The CRAPome: a contaminant repository for affinity purification mass spectrometry data. Nat Methods 10:730–736

29. Doaga IO, Savopol T, Neagu M et al (2008) The kinetics of cell adhesion to solid scaffolds: an experimental and theoretical approach. J Biol Phys 34:495–509

Chapter 15

Examining the Effects of Nuclear GPCRs on Gene Expression Using Isolated Nuclei

George Vaniotis, Sarah Gora, André Nantel, Terence E. Hébert, and Bruce G. Allen

Abstract

The measurement of changes in the transcriptome is a common end point for various pathologic and pharmacologic studies. In recent years, with the discovery of a host of potential pharmacologic targets located directly on the nuclear membrane, the need to assess their potential control over the transcriptome has arisen. Here we present techniques for assessing changes in gene expression in isolated nuclei in response to stimulation by endogenous GPCRs on the nuclear membrane.

Key words Intracrine signaling, Transcriptome, qPCR, Nucleus, Gene, RNA

1 Introduction

The measurement of changes, both positive and negative, at the level of transcription has been widely used as a tool to assess the differences produced by various pathologies as well as the effects of various treatments either in the whole animal or in individual cells. By assessing how the transcription of certain genes is altered, we can ascertain what pathways are being modulated as well as what physiologic effects might be expected as a result. The control of gene transcription is a complex process involving a variety of different receptors, protein kinases, transcription factors as well as a host of other factors including RNA polymerases [1]. Recently, evidence has begun to accumulate demonstrating a pool of G protein-coupled receptors (GPCRs), as well as their signaling partners located in the nuclear envelope and in the nucleus [2]. Given that these receptors are located on the nuclear membrane, they are ideally placed to play a role in regulating gene expression and, perhaps, chromatin remodeling. As such, a way to assess whether these receptors might play a role in regulating gene transcription is essential in determining their physiological function.

Bruce G. Allen and Terence E. Hébert (eds.), *Nuclear G-Protein Coupled Receptors*, Methods in Molecular Biology, vol. 1234, DOI 10.1007/978-1-4939-1755-6_15, © Springer Science+Business Media New York 2015

Here we describe a methodology to examine transcriptional events in isolated nuclei. The first method, a transcription initiation assay, allows for the measurement of changes in global transcription. This approach provides benefits as well as drawbacks. By virtue of measuring global transcription, it is possible to ascertain whether transcription, as a whole, is increased or decreased across the entire genome. However this also means that if the effect of the treatment results in both up- and downregulation of groups of genes, cumulative effects might not be visible, thus limiting its discriminatory power. In addition, the analysis could be biased if the transcription of high abundance genes, such as ribosomal RNA, is affected. Another widely used technique is quantitative PCR (qPCR) [3]. In contrast to the global portrait provided by the transcription initiation, qPCR generates data on the RNA levels of specific genes. This approach is useful as long as there is some idea of which genes to look at. However, this approach is not suited for studying the effects of intracrine signaling on transcription in a broader context. As such, it is generally necessary to first employ either a pathway-targeted or a genome-wide microarray approach to identify potential sets of genes that can then be validated by qPCR. A technique that is beginning to be more widely used is RNA sequencing, which permits comprehensive characterization of the whole transcriptome [4]. RNA sequencing has several advantages over other expression profiling technologies including higher sensitivity and the ability to detect splice variants [5].

2 Materials

Prepare all solutions using Grade 1 (ultrapure; dH$_2$O) water and use analytical grade reagents wherever available.

2.1 Components

GF/C glass microfiber filters (24 mm diameter).

Filtration manifold.

Mortar and pestle.

Liquid nitrogen and Dewar.

Metal Spatula.

Ice bucket.

Quantitative PCR system.

Refrigerated microcentrifuge.

2.2 Buffers

1. TRIS-buffered saline (TBS): Dissolve 6.05 g TRIS base and 8.76 g NaCl in 800 mL of dH$_2$O. Adjust pH to 7.8–7.9 with 1 M HCl. The pH of TRIS solutions is highly temperature dependent: ensure solution is at room temperature. If not, allow to cool. Adjust pH to 7.8 with 1 M HCl. Adjust volume to 1 L with dH$_2$O. Store at 4 °C for up to 3 months.

2. TBS Buffer: Add 10 μL of 1.0 M DTT, 20 μL of 0.20 M Na$_3$VO$_4$, and 25 μL of 0.20 PMSF to 18 mL of TBS.

3. Isolation Buffer A: Weigh out 238.4 mg of HEPES into a glass beaker. Add dH$_2$O to a volume of 80 mL. Add 150 μL of 1.0 M MgCl$_2$, 50 μL of 1.0 M DTT, 2.5 mg of leupeptin, 100 μL of 0.20 M Na$_3$VO$_4$, and 333 μL of 3.0 M KCl. Mix using a stir bar until the HEPES is fully dissolved. Adjust pH to 7.8–7.9 with 1 M HCl. Transfer to a graduated cylinder and complete to 100 mL with dH$_2$O. Store at 4 °C.

4. Isolation Buffer B: Weigh out 7.152 g of HEPES and 10.430 g of KCl into a glass beaker. Add dH$_2$O to a volume of 80 mL. Add 3 mL of 1.0 M MgCl$_2$, 2.5 mg of leupeptin, and 100 μL of 0.20 M Na$_3$VO$_4$. Mix using a stir bar until the HEPES and KCl are fully dissolved. Adjust pH to 7.8–7.9 with 1 M HCl. Transfer to a graduated cylinder and complete to 100 mL with dH$_2$O. Store at 4 °C.

5. Isolation Buffer C: Weigh out 476.8 mg of HEPES, 3.155 g of glycerol, and 2.450 g of NaCl into a glass beaker. Add dH$_2$O to a volume of 80 mL. Add 150 μL of 1.0 M MgCl$_2$, 50 μL of 1.0 M DTT, 2.5 mg of leupeptin, 5.8 mg of EDTA, 100 μL of 0.20 M Na$_3$VO$_4$, and 250 μL of 0.20 M PMSF. Mix using a stir bar until the HEPES, glycerol, and NaCl are fully dissolved. Adjust pH to 7.8–7.9 with 1 M HCl. Transfer to a graduated cylinder and complete to 100 mL. Store at 4 °C.

6. Transcription Buffer (3×): In a 1.5 mL microcentrifuge tube combine 150 μL 1.0 M Tris–HCl, pH 7.9 (RT), 150 μL 3.0 M KCl, 10 μL 0.30 M MnCl$_2$, 18 μL 1.0 M MgCl$_2$, 100 μL 30 mM ATP, 6 μL 1.0 M DTT, and 75 μl 40 U/μL RNAse inhibitor. Complete to 1 mL with dH$_2$O and mix.

7. Transcription Lysis Buffer: 10 mM Tris–HCl, pH 8.0 (RT), 10 mM EDTA, and 1 % SDS. Complete with dH$_2$O.

8. Filter Blocking Solution: 25 mM Tris–HCl, 7.4, 0.15 % BSA, and 0.3 % polyethylenimide. Complete with dH$_2$O, store at 4 °C.

3 Methods

Carry out all procedures at room temperature unless otherwise specified.

3.1 Rat Heart Extraction

1. Anesthetize animal using Somnotol (9.2 mg/100 g).

2. Ensure the animal is properly anesthetized by pinching its paws with a set of tweezers.

3. Surgically remove tissue of interest (heart) and snap freeze in liquid nitrogen to preserve the heart (*see* **Note 1**).

3.2 Tissue Homogenization

1. Fill the ice bucket with ice and place the mortar in the middle.

2. Pour liquid nitrogen over the ice and into the mortar in order to cool the mortar and the surrounding ice.

3. Try to ensure that the mortar is always approximately half full of liquid nitrogen (*see* **Note 2**).

4. Transfer the heart(s) from the liquid nitrogen container to the mortar.

5. Slowly pulverize the heart(s) with the pestle into a fine powder (*see* **Note 3**).

6. Chill a metal spatula in liquid nitrogen and then use it to transfer the powdered tissue to a 50 mL Falcon tube containing 18 mL TBS buffer with the spatula (*see* **Note 4**).

7. Homogenize the above solution using a Polytron homogenizer on low, for 20 s.

8. Collect a 1 mL aliquot. Label this Fraction A.

3.3 Nuclear Isolation

1. Centrifuge homogenate at $500 \times g$ for 15 min at 4 °C.

2. Transfer supernatant to a fresh 50 mL Falcon tube.

3. Collect a 1 mL aliquot. Label this Fraction B.

4. Add 1 volume of Isolation Buffer A, 1 mL at a time, while gently mixing.

5. Incubate on ice for 10 min.

6. Centrifuge at $2,000 \times g$ for 15 min at 4 °C.

7. Discard supernatant.

8. Resuspend the pellet from **step 6** in 2 mL of Isolation Buffer B, mix gently until solubilization, and add another 3 mL of Isolation Buffer B.

9. Mix gently and incubate on ice for 20 min.

10. Collect a 1 mL aliquot. Label this Fraction C.

11. Centrifuge at $2,000 \times g$ for 15 min at 4 °C.

12. Discard the supernatant.

13. The buffer employed to resuspend the pellet from **step 11** depends upon what the nuclei will be used for. If the nuclei are to be employed in studies of transcription, proceed directly to Subheading 3.4 (global assessment of transcription using the transcription initiation assay) or Subheading 3.5 (gene-targeted assessment of transcription). Otherwise suspend the pellet in 1 mL of Isolation Buffer C and label this as Fraction D.

3.4   Assessment of Global Transcription

This assay can be used to assess the effect of a specific ligand upon global transcription in isolated nuclei but will not provide information about the expression of individual genes and the values obtained in this manner will be a summation of both positive and negative effects on gene expression.

3.4.1   Transcription Initiation Assay

1. Following the preparation of isolated nuclei from adult rat hearts as described in Subheading 3.3, resuspend the pellet obtained in Subheading 3.3, **step 11** using 1 mL of Transcription Buffer (1×).

2. Place the GF/C filter disks in filter blocking solution. Soak for at least 1 h

3. Prepare the Transcription Reaction Media: to 3× Transcription Buffer add [α^{32}P]UTP to a final concentration of 1 μCi/μL (*see* **Note 5**).

4. Prepare tubes in triplicate, containing 10 μL of the drug(s) of interest, 10 μL of the isolated nuclei, and 10 μL of the Transcription Buffer (with [α^{32}P]UTP).

5. Incubate at 37 °C for 30 min (*see* **Note 6**).

6. Terminate the reaction by adding DNAse I to each tube to a final concentration of 0.1 U/μL.

7. Incubate at 37 °C for 30 min.

8. Add 30 μL of Transcription Lysis Buffer to each tube.

9. Incubate on ice for 30 min.

10. Prepare the filtration system. This includes placing treated GF/C filter disks into the filtration manifold and applying vacuum.

11. Transfer 55 μL aliquots of the reaction mixture onto the GF/C filters one at a time. Apply to centre of filter. Immediately after applying sample onto the filter, rinse with 5 % (w/v) TCA solution containing 20 mM sodium glycerophosphate.

12. Remove the filter discs from the filtration manifold and place them into scintillation vials.

13. Add 5 mL of Fluor to each vial and determine the ^{32}P incorporated by liquid scintillation counting.

14. Use the remaining isolated nuclei to assess the DNA concentration (as detailed in Subheading 3.5). Normalize ^{32}P (cpm) incorporation to DNA content (μg).

3.4.2   DNA Extraction

1. Transfer 300 μL of isolated nuclei, from Subheading 3.4, **step 14**, into a 1.5 mL microcentrifuge tube. Add 30 μg of proteinase K and adjust to 0.5 % SDS.

2. Incubate overnight at 50 °C.

3. The following morning, incubate at 80 °C for 30 min to inactivate the proteinase K.

4. Add 50 µg of RNAse A, incubate at 37 °C for 1 h.

5. Add 400 µL of phenol, mix by manually inverting the tube for 5 min.

6. In a microcentrifuge, centrifuge for 5 min at $13,800 \times g$ (13,000 rpm, Eppendorf 5415C centrifuge with 18-place rotor) at room temperature. Recover the upper aqueous layer using a 1 mL pipette.

7. Add 200 µL of phenol and 200 µL of chloroform. Mix for 1 min.

8. Centrifuge for 5 min at $13,800 \times g$ (13,000 rpm, Eppendorf 5415C centrifuge with 18-place rotor) at room temperature, recover upper aqueous layer, using a 1 mL pipette.

9. Add 400 µL of chloroform. Mix 30 s.

10. Centrifuge for 30 s at $13,800 \times g$ (13,000 rpm, Eppendorf 5415C centrifuge with 18-place rotor) at room temperature. Recover the upper aqueous layer using a 1 mL pipette.

11. Add 1 volume of 5 M sodium acetate. Mix by inversion.

12. Add 2 volumes of 100 % ethanol. Mix by inversion.

13. Incubate overnight at –20 °C to precipitate DNA.

14. Centrifuge for 15 min at $18,300 \times g$ (14,000 rpm, Thermo IEC Micromax RF centrifuge) at 4 °C.

15. Wash the pellet with 200 µL of 70 % ethanol (*see* **Note 7**).

16. Centrifuge for 15 min at $18,300 \times g$ (14,000 rpm, Thermo IEC Micromax RF centrifuge) at 4 °C. Decant the supernatant.

17. Remove all of the liquid from the pellet. Then let air-dry for approximately 15 min.

18. Resuspend the DNA in 20 µL of sterile dH$_2$O. Quantify (*see* **Note 8**).

19. Express results from transcription initiation assays by normalizing ^{32}P (cpm) incorporation to DNA content (µg) for each assay (Fig. 1a).

3.5 Gene-Targeted Assessment of Transcription

The techniques described in this section may be used to assess the effect of ligands upon the transcription of one or more specific genes.

3.5.1 Stimulation of Isolated Nuclei

1. Following preparation of isolated nuclei from adult rat hearts as described in Subheading 3.3, resuspend the pellet obtained in Subheading 3.3, **step 11** using 1 mL of Isolation Buffer C.

2. Split the sample into separate tubes, the number of which will vary depending on your desired conditions (*see* **Note 9**).

3. Add the desired drug(s) to the appropriate tubes, taking into account the final volume of the treatment concentration.

4. Incubate at 37 °C for 30 min (or for different times as required).

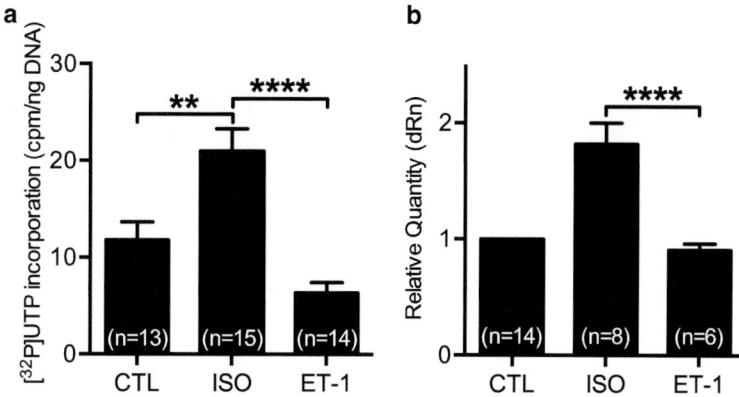

Fig. 1 Assessment of transcription initiation and the expression of ribosomal RNA in isolated nuclei. Nuclei isolated from four adult rat hearts were treated with isoproterenol (ISO, 1 μM), endothelin-1 (ET-1, 10 nM), or vehicle for 30 min at 37 °C. Transcription initiation (**a**) and 18S ribosomal RNA (**b**) were quantified as described in Subheadings 3.4.1–3.4.2 and 3.5.1–3.5.2, respectively. The C_T values for 18S rRNA were normalized to those for β-actin mRNA. The primers used for 18S and β-actin mRNA have been described previously [6]. Data are presented as the mean ± the standard error of the mean (s.e.m.). The number of nuclear preparations analyzed is indicated in parentheses. The significance of differences between groups was determined using one-way ANOVA followed by Tukey's Multiple Comparison test (GraphPad Prism version 6.00 for Mac OS X, GraphPad Software). **, $P < 0.01$; ****, $P < 0.0001$

3.5.2 RNA Isolation/
cDNA Generation

1. Add 1 mL of Tri Reagent to nuclei obtained at the end of Subheading 3.3, let sit at room temperature for 5 min (*see* **Note 10**).

2. Add 200 μL of chloroform to each sample. Vortex. Let sit at room temperature for 3 min.

3. Centrifuge for 15 min at 14,000 rpm (18,300 × *g*, Thermo IEC Micromax RF centrifuge) at 4 °C.

4. Collect the upper layer using a 1 mL pipette. Add 1 volume of 70 % ethanol.

5. Isolate RNA using "Qiagen" RNA isolation RNeasy Mini kit.

6. Elute using 25 μL of sterile RNAse-free water

7. Quantify RNA using a spectrophotometer, or preferably a Nanodrop, paying particular attention to the 260/280 ratio (Table 1; *see* **Note 11**).

8. Add in a tube 100 ng random primers, 1 μg of RNA, 1 μL of 10 mM dNTP. Mix and complete to 17 μL with RNAse-free water.

9. Heat mixture to 65 °C for 5 min.

10. Add 4 μL of 5× first strand buffer (FSB), 2 μL of 0.1 M DTT, and 1 μL of RNAse inhibitor.

Table 1
RNA quality analysis using a Bioanalyzer

Bioanalysis RNA	Concentration	A260/A280	A260/A230	Ratio 28S/18S	RIN	Quality
Total Heart	362.61 ng/µL	1.92	2.43	1.955	9.5	Good
Nuclear Heart 1	207.19 ng/µL	2.03	2.39	1.300	3.8	Bad
Nuclear Heart 2	295.90 ng/µL	2.03	2.22	1.187	3.6	Bad

11. Mix contents by hand, centrifuge, and incubate at 37 °C for 2 min.

12. Add 1 µL of M-MLVRT, mix by hand, and centrifuge.

13. Incubate at 25 °C for 10 min, then at 37 °C for 50 min, and finally at 70 °C for 15 min.

14. cDNA can be stored at –20 °C, in aliquots, if desired.

3.5.3 qPCR

Levels of individual transcripts may be quantified, relative to that of one or more "house-keeping" genes, by real-time quantitative PCR (qPCR).

1. Add in a tube 12.5 µL of SyBr Green with ROX (*see* **Note 12**), 10 µL of cDNA, and 2.5 µL of a 0.3 µM dilution of the desired primers.

2. Primer sequences can either be obtained from the literature or designed in-house using software designed for this regard, as well as Primer-Blast at the NCBI website (*see* **Note 13**).

3. Primers for each gene of interest should be validated generating a dilution curve, to ascertain their linear range, and a dissociation curve to ensure only single products are being amplified. Sequencing the amplicon is also advisable.

4. Each sample should be analyzed in duplicate, or even triplicate and a tube with no cDNA should also be included as a control, referred to as a no template control (NTC).

5. Expression levels for housekeeping genes should also be determined for each sample. These will serve as internal standards (*see* **Note 14**).

6. Following amplification and analysis, expression levels of genes of interest should be normalized to those of the housekeeping genes and the calibrator, usually the vehicle-treated control (Fig. 1b).

3.5.4 PCR Array/ Gene Chip

Similarly to qPCR, genome-wide or targeted gene arrays can be employed, once again using cDNA generated from the desired samples to generate a more general assessment of changes in gene

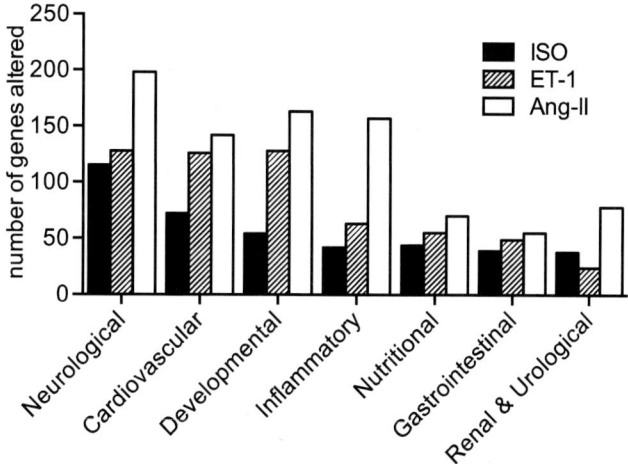

Fig. 2 Classification of the effects of isoproterenol, endothelin-1, and angiotensin-II on transcription in isolated nuclei according to pathology. Nuclei isolated from four adult rat hearts were treated with isoproterenol (ISO, 1 μM), endothelin-1 (ET-1, 10 nM), angiotensin-II (AngII, 10 nM), or vehicle for 30 min at 37 °C. RNA was then isolated and gene expression analysis was performed using Agilent Rat Whole Genome microarrays. Data was analyzed using Ingenuity Systems software and presented as a ratio of expression versus vehicle control. Total rat RNA was used to normalize the results. The results shown comprise the mean of three preparations

expression. In the case of the targeted PCR arrays, pre-prepared plates containing primers against a specific pathway or set of genes are used [6]. Any genes that appear to change in these arrays can be further confirmed by qPCR. Using microarrays or RNA sequencing, the entire genome can be probed, allowing for assessment of global changes in transcription resulting from activation of receptors on the nuclear envelope. These techniques, however, require the use of specialized software that is capable of analyzing the large amount of data that ensues (Fig. 2), such as GeneSpring or Ingenuity. Again, in the case where microarrays are employed, key individual results must be validated by qPCR.

4   Notes

1. Multiple hearts can be pooled in order to ensure enough material is obtained to conduct the experiment.

2. Ensure that the heart(s) remain completely immersed in liquid nitrogen during this step.

3. Start by breaking the heart(s) into smaller pieces that can then be more easily ground down. Ensure not to use too much force as this can cause splashing of the liquid nitrogen and loss of material.

4. Place the spatula in the liquid nitrogen container in order to cool it before use.

5. Assure that you account for the half-life of the radioisotope (14.28 days for ^{32}P).

6. At this time also place an appropriate number of glass fiber filter circles in the fixer solution, for at least an hour before use.

7. Simply invert tube a few times, pellet should not dissolve.

8. Quantify DNA spectrophotometrically. Verify that the 260/280 ratio is above 1.8. An absorbance unit at 260 nm corresponded to 50 μg/mL of DNA. If available, use a microspectrophotometer (i.e., NanoDrop) that permits assays using 1–2 μL of sample.

9. The nuclei isolated from two rat hearts must not be divided into more than 6 separate transcription reactions to ensure the yield is sufficient for subsequent analysis.

10. If isolating RNA from cardiomyocytes, rather than heart tissue, centrifuge sample at $82 \times g$ (1,000 rpm, Eppendorf 5415C centrifuge with 18-place rotor) at room temperature for 5 min first, then resuspend pellet in 1 mL of Tri Reagent and homogenize by Polytron for 20 s, prior to incubating at room temperature for 5 min.

11. If using a Bioanalyzer to assess RNA quality, it should be noted that it is expected that the ribosomal RNA ratio will be off, leading to a false assessment of low quality RNA. This is due to the fact that most rRNA present in the cytoplasm is lost during the process of isolating the nucleus, thus altering the expected ratio.

12. Add 3 μL of ROX for every 1,250 μL of SyBr Green.

13. When designing primers, a few key points to keep in mind are primer GC content should be between 50 % and 60 %, primer melting temperature (Tm) should be between 55 °C and 65 °C, amplicon size should ideally range from 75 to 300 nucleotides, the primer itself should be about 20 nucleotides long, and should recognize the gene in your organism of interest.

14. Make sure to select a proper housekeeping gene, by verifying that its expression does not change with your treatment of choice. Potential housekeeping genes include GAPDH, β-actin, and 18S rRNA but these must be generated empirically, based on results obtained.

Acknowledgements

This work was supported by grants from the Canadian Institutes of Health Research (MOP-64183 and MOP-125970 to BGA and MOP-79354 and MOP-119530 to TEH) and the Fondation des Maladies du Coeur du Québec (to BA).

References

1. Bywater MJ, Pearson RB, McArthur GA et al (2013) Dysregulation of the basal RNA polymerase transcription apparatus in cancer. Nat Rev Cancer 13:299–314

2. Tadevosyan A, Vaniotis G, Allen BG et al (2012) G protein-coupled receptor signalling in the cardiac nuclear membrane: evidence and possible roles in physiological and pathophysiological function. J Physiol 590:1313–1330

3. Zhu T, Gobeil F, Vazquez-Tello A et al (2006) Intracrine signaling through lipid mediators and their cognate nuclear G-protein-coupled receptors: a paradigm based on PGE2, PAF, and LPA1 receptors. Can J Physiol Pharmacol 84:377–391

4. Loewe RP (2013) Combinational usage of next generation sequencing and qPCR for the analysis of tumor samples. Methods 59:126–131

5. Huang Q, Lin B, Liu H et al (2011) RNA-Seq analyses generate comprehensive transcriptomic landscape and reveal complex transcript patterns in hepatocellular carcinoma. PLoS One 6:e26168

6. Vaniotis G, Del Duca D, Trieu P et al (2011) Nuclear β-adrenergic receptors modulate gene expression in adult rat heart. Cell Signal 23:89–98

Chapter 16

Trafficking and Function of GPCRs in the Endosomal Compartment

Davide Calebiro, Amod Godbole, Sandra Lyga, and Martin J. Lohse

Abstract

New methods based on fluorescently labeled agonists, genetically encoded fluorescent sensors, and advanced microscopy techniques, such as fluorescence resonance energy transfer (FRET) and highly inclined thin illumination (HILO), allow direct monitoring of signaling, internalization, and intracellular trafficking of G protein-coupled receptors (GPCRs) and their ligands in living cells with high temporal and spatial resolution. These methods have been essential in revealing that GPCRs can continue signaling via production of the soluble second messenger cyclic AMP after internalization into the endosomal compartment.

Key words Fluorescence resonance energy transfer, Total internal reflection fluorescence microscopy, Highly inclined thin illumination, G protein-coupled receptor, Thyroid stimulating hormone, Thyroid stimulating hormone receptor, Endocytosis inhibitor, Thyroid follicle

1 Introduction

G protein-coupled receptors (GPCRs) have traditionally been investigated using pharmacological (e.g., radioligand binding assays) and biochemical (e.g., radioimmunoassays) methods on cell lysates and purified membranes. Recent advances in chemistry, protein engineering, and optics have brought about the possibility of directly observing GPCR signaling as it occurs in living cells and tissues [1]. Indeed, whereas GPCRs have long been believed to signal via G proteins only when located on the cell surface, recent studies, made possible by the use of genetically encoded fluorescent sensors and probes, have shed new light on the function of receptors after internalization into endosomes, which are emerging as previously overlooked, intracellular platforms for GPCR signaling [2–7].

This chapter describes: (1) How to visualize the binding of a fluorescent hormone, i.e., the thyroid stimulating hormone (TSH), to its endogenous GPCR located on the surface of thyroid cells and follow its internalization and trafficking in real time by highly

Bruce G. Allen and Terence E. Hébert (eds.), *Nuclear G-Protein Coupled Receptors*, Methods in Molecular Biology, vol. 1234, DOI 10.1007/978-1-4939-1755-6_16, © Springer Science+Business Media New York 2015

inclined thin illumination (HILO) microscopy; (2) How to use fluorescence resonance energy transfer (FRET) microscopy to monitor the cAMP response to TSH hormone stimulation in primary thyroid cells expressing a fluorescent biosensor for cAMP; (3) How to use these methods together with a pharmacological inhibitor of clathrin-dependent endocytosis to reveal the occurrence of GPCR-cAMP signaling on endosomal membranes after internalization of receptors together with their ligands. Whereas the protocols described in this chapter have been optimized to study TSH receptor signaling in thyroid cells, they can be adapted to monitor the trafficking and signaling of other GPCRs in various cellular models, including cell lines, primary cells, and tissue slices [2, 8, 9].

2 Materials

Prepare all solutions with ultrapure water (obtained by filtering deionized water to achieve a resistance of 18 MΩ at 25 °C) or, when specifically indicated, with cell culture-grade dimethyl sulfoxide (DMSO). Common lab equipment, including pipette aids, sterile pipettes and tips, and a bench centrifuge, as well as equipment for cell culture, such as an incubator and a laminar flow hood, are required. Handle cells and cell culture reagents/materials under sterile conditions, while working under a laminar flow hood.

2.1 Isolation of Mouse Primary Thyroid Follicles and Thyrocyte Culture

1. Dulbecco's phosphate buffered saline (DPBS): Without Ca^{2+} and Mg^{2+}, sterile filtered. Store at 4 °C.

2. Hank's balanced salt solution (HBSS): Sterile filtered. Store at 4 °C.

3. Phenol red-free Dulbecco's modified Eagle's medium (DMEM)/F-12 medium: Sterile filtered. Store at 4 °C.

4. Complete culture medium: Prepare by adding 20 % (v/v) fetal bovine serum (FBS), 100 U/mL of penicillin, and 100 μg/mL of streptomycin to phenol red-free (see **Note 1**) DMEM/F-12 medium. Store at 4 °C.

5. Trypsin/EDTA solution: 0.05 % (w/v) Trypsin and 0.02 % (w/v) EDTA in DPBS without Ca^{2+} and Mg^{2+}, sterile filtered. Store at 4 °C.

6. 0.22-μm sterile filters.

7. Collagenase I (Gibco): Prepare 1,000 U/mL stock solution in HBSS, aliquot and store at −20 °C.

8. Collagenase II (Gibco): Prepare 1,000 U/mL stock solution in HBSS, aliquot and store at −20 °C.

9. Dispase II (Roche Diagnostics): Prepare 10 U/mL stock solution in HBSS, aliquot and store at −20 °C.

10. Collagen (Roche diagnostics): 3 mg/mL in 0.2 % acetic acid. Prepare a 0.2 % acetic acid solution and sterilize it by passing through a 0.22-μm sterile filter. Add the solution to the lyophilized collagen. Do not mix or vortex but rather let dissolve O/N at room temperature and store at 4 °C.

11. Neutralization buffer: 0.4 M NaHCO$_3$, 0.2 M 4-(2-hydroxyethyl)-1-piperazineethanesulfonic acid (HEPES), pH 7.4. Sterilize with a 0.22-μm filter under a laminar hood and store at 4 °C.

12. 24-mm round-glass coverslips (0.13–0.16-mm thick): Sterilize by immersing in 100 % ethanol. Wash briefly with DPBS before use.

13. 6-well cell culture plates and 10-cm Petri cell culture dishes.

14. 35-mm glass-bottom cell culture dishes (World Precision Instruments).

15. Collagen gel-coated glass-bottom Petri dishes: Place 8 μL of the collagen solution in the center of a glass-bottom cell culture dish and spread it to uniformly cover the coverslip surface (*see* **Note 2**). Add 200 μL neutralization buffer to cover the collagen. Place in an incubator at 37 °C for 30–45 min. Wash the dish with 2 mL of DPBS and store for up to 12 h with DPBS at 4 °C.

16. Dissecting board and pins.

17. Stereomicroscope or magnifying glass with support.

18. Dissection instruments: Pair of broad-pointed forceps, two pairs of fine forceps (straight and curved), pair of large scissors, pair of ultrafine scissors (*see* **Note 3**). Autoclave before use.

19. Water bath maintained at 37 °C.

2.2 Preparation of Fluorescently Labeled Thyroid Stimulating Hormone (See Note 4)

1. Alexa-Fluor dye, succinimidyl ester (Invitrogen) with the desired fluorescence excitation and emission spectra (*see* **Note 5**): 1 mg vial. Store at −20 °C.

2. Bovine TSH (Sigma-Aldrich): 5 mg/vial. Store at 4 °C.

3. Dimethyl sulfoxide (DMSO).

4. 0.1 M NaHCO$_3$ solution: prepare fresh.

5. Column wash solution: 20 mM NaCl, 20 mM Tris–HCl, pH 7.4. Prepare fresh.

6. Empty chromatography column (*see* **Note 6**), vertical stand with clamps, tubings and fittings, reservoir for the wash buffer, online UV (280 nm) spectrophotometer, and fraction collector.

7. Sephadex G-25 slurry: Suspend 5 g of normal quality Sephadex G-25 (GE Healthcare Life Sciences) in 50 mL of column wash solution. Allow Sephadex G-25 to swell. Store at 4 °C.

8. DPBS without Ca^{2+} and Mg^{2+}.

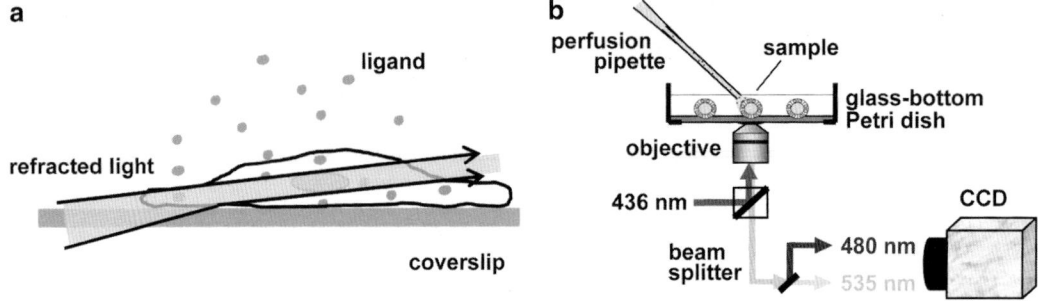

Fig. 1 Schematic showing principles of HILO (**a**) and FRET (**b**) microscopy

2.3 Visualization of Ligand:Receptor Complex Internalization and Trafficking by HILO Microscopy

1. Total internal reflection fluorescence (TIRF) microscope: Inverted fluorescence microscope, equipped with an oil-immersion high numerical aperture objective (e.g., 100× magnification/1.46 numerical aperture), suitable lasers (e.g., 405, 488, 561, and 645-nm diode lasers) and filters, TIRF illumination, an electron multiplying charge coupled device (EMCCD) camera, and an incubator with a temperature control set at 37 °C (Fig. 1a) (*see* **Note 7**).

2. 0.22-μm sterile filters.

3. Bovine serum albumin Fraction V (BSA).

4. Imaging medium: Supplement an aliquot of complete culture medium with 1 % (w/v) BSA (*see* **Note 8**). Sterilize with a 0.22-μm filter. Prepare fresh for each experiment.

5. Dynasore stock solution: dissolve dynasore (Sigma-Aldrich) in DMSO under a sterile hood to a final concentration of 40 mM. Store at –20 °C.

6. Fluorescently labeled TSH.

7. Imaging chamber (e.g., Attofluor Cell Chamber, Molecular Probes).

8. Wild-type mice: male or female, aged 1–6 months.

2.4 Transfection of Mouse Primary Thyroid Cells by Electroporation

1. One 10-cm Petri dish of semi-confluent mouse primary thyroid cells.

2. DPBS.

3. Trypsin/EDTA solution.

4. Sterile 24-mm round-glass coverslips (0.13–0.16-mm thick).

5. 6-well cell culture plate.

6. 15-mL sterile plastic tubes with cap.

7. Centrifuge.

8. Water bath maintained at 37 °C.

9. Cuvettes for electroporation (e.g., 0.4-cm electrode gap cuvettes, Bio-Rad).

10. Electroporator (e.g., Gene Pulser with Capacitance Extender, Bio-Rad).

11. Epac1-camps (*see* **Note 9**) [10] plasmid.

2.5 Real-Time Monitoring of cAMP Signaling by FRET Microscopy

1. Fluorescence resonance energy transfer (FRET) microscope (*see* **Note 10**): Inverted fluorescence microscope, equipped with a 10× (air) objective and a high-magnification oil-immersion objective (e.g., 63×/1.25 numerical aperture), a stable light source (e.g., xenon lamp, xenon lamp with monochromator or 435-nm light-emitting diode), appropriate filters, i.e., 436/20 (excitation) plus 455LP (dichroic) for CFP excitation and 500/20 (excitation) plus 515LP (dichroic) for YFP excitation, a beam splitter (e.g., DualView, Photometrics) with a 505LP dichroic mirror and emission filters for CFP (480/30) and YFP (535/40), an EMCCD camera, an incubator, and a temperature control set at 37 °C (Fig. 1b).

2. Software for controlling the microscope and image acquisition (e.g., MetaFluor, Molecular Devices) on a PC connected to the FRET microscope.

3. 0.22-μm sterile filters.

4. Bovine serum albumin Fraction V (BSA).

5. Imaging medium: Supplement an aliquot of complete culture medium with 1 % (w/v) BSA (*see* **Note 11**). Sterilize with a 0.22-μm filter. Prepare fresh for each experiment.

6. Bovine TSH (Sigma-Aldrich): 5 mg/vial, stored at 4 °C. Prepare stock solution by dissolving in DMEM/F12 medium at a concentration of 2,000 U/L. Make aliquots and store at −20 °C.

7. Imaging chamber (e.g., Attofluor Cell Chamber, Molecular Probes).

8. Electronic spreadsheet for calculation of corrected FRET values.

9. Multichannel perfusion system (e.g., OctaFlow II, ALA Scientific Instruments).

10. Micromanipulator with shaft: Attach the tip of the perfusion system to the shaft of the micromanipulator.

11. Aspirator with suction tubing to remove the excess liquid from the chamber.

3 Methods

Carry out all procedures at room temperature unless otherwise specified.

3.1 Isolation of Mouse Primary Thyroid Follicles and Thyrocyte Culture

1. Work under aseptic conditions to avoid contamination in primary cell cultures. Use designated places and extensively cleaned spaces for dissections and always employ autoclaved dissection instruments.

2. Prepare 1 mL of digestion medium by adding 100 μL each of collagenase I, collagenase II, and dispase II stock solutions to 700 μL DMEM/F12 medium without FBS and antibiotics in a sterile tube with cap.

3. Sacrifice the mouse by cervical dislocation and sterilize the skin (*see* **Note 12**).

4. Fix the mouse on the dissecting board by pinning the palms, feet, and snout so that the animal lies in a supine position with the throat arched upwards.

5. Make a vertical incision just below the throat with the help of a broad-pointed forceps and large scissors and extend it upwards and sideways to expose the inner organs.

6. With the help of fine forceps, remove the salivary glands. From this step on use a stereomicroscope or magnifying glass. Isolate the muscles covering the trachea with the help of the fine curved forceps and cut away the muscles with the ultrafine scissors. Free the trachea from the esophagus using the curved forceps and cut it just above the sternum with the help of the ultrafine scissors. Gently lift the trachea with fine forceps and bend it upwards to reveal the two thyroid lobes on either side.

7. Carefully remove both thyroid lobes with the help of the two fine forceps (*see* **Note 13**). Briefly wash them by dipping in sterile DPBS and transfer them into the tube containing the digestion medium. Repeat **steps 3–7** with the remaining mice (*see* **Note 14**).

8. Carry out the enzymatic digestion at 37 °C in a water bath incubator for 90 min, while gently inverting the tube every 15 min (*see* **Note 15**).

9. Wash the digested follicles thrice by adding 10 mL of complete culture medium and centrifuging at $100 \times g$ each time.

10. To work with intact follicles, plate the follicles on collagen gel-coated coverslips and incubate at 37 °C, 5 % CO_2 for 2–3 h to let the follicles adhere to the collagen gel. Perform microscopy experiments 3–12 h after plating.

11. To obtain individual thyroid cells, isolate follicles from three mice, wash them as described above, and seed them directly in a 10-cm Petri culture dish. Culture at 37 °C, 5 % CO_2 for 7–10 days changing the medium every 3–4 days. Thereafter, replate the cells on 24-mm round coverslips by following this procedure: Place 24-mm round glass coverslips in two 6-well

plates. Pre-warm DPBS and the trypsin/EDTA solution at 37 °C, wash the cells once with 10 mL pre-warmed DPBS, add 2 mL trypsin/EDTA solution, incubate for 10 min at 37 °C, 5 % CO_2 in an incubator. Tap the Petri dish with the palm of a hand to fully detach the cells. Collect the cells by adding 10 mL of complete culture medium and transfer to a sterile plastic vial. Bring the volume to 25 mL with additional complete culture medium, mix by gently pipetting up and down. Dispense 2 mL of the medium with thyroid cells to each well of the 6-well plate. Incubate at 37 °C, 5 % CO_2 for 24 h in an incubator before use.

3.2 Preparation of Fluorescently Labeled Thyroid Stimulating Hormone

1. Carry out all the steps on ice/in a cold room unless otherwise indicated.

2. Fix the empty chromatography column to a vertical stand using clamps.

3. Clean the column with ultrapure water.

4. Fill the column up to one-quarter of its height with column wash solution.

5. Add the suspended Sephadex G-25 and allow it to settle into a compact form. Remove the solution from the top using a Pasteur pipette. Repeat this procedure until the column is filled with Sephadex G-25, leaving the space for approx. 2 mL of wash solution above the Sephadex G-25.

6. Connect the column with a tubing to a reservoir containing the column wash buffer placed on a higher position so that the solution can flow by gravity. Wash the column by running 100 mL of column wash solution. Take care that the column does not run dry (*see* **Note 16**).

7. Take two vials of bovine TSH, 5 mg each, dissolve them in 0.5 mL ice-cold DPBS, and transfer the solution to a 1.5-mL plastic tube with cap.

8. Take the vial containing 1 mg Alexa-Fluor reactive dye and wrap it in aluminum foil to protect it from light. Add 50 μL DMSO to the vial to dissolve the dye. Vortex it shortly to completely dissolve the dye (*see* **Note 17**).

9. Add 50 μL of 0.1 M $NaHCO_3$ solution to the dissolved TSH. Gently mix and add the dissolved Alexa-Fluor dye, close the vial, and cover it completely with aluminum foil. Mix by inverting the vial 3–5 times.

10. Attach the vial to a tube rotator and incubate for 90 min at room temperature under slow rotation.

11. Remove the upper cap of the Sephadex-G25 column, load the labeled TSH solution, and reclose the column.

12. Elute the column with column wash buffer. Use the fraction collector to obtain several fractions while recording a chromatogram with the UV spectrophotometer. Two well-separated peaks should be visible on the chromatogram. Combine all the fractions corresponding to the first peak (i.e., labeled TSH) and discard those corresponding to the second peak (free unreacted dye).

13. Aliquot the labeled TSH, protect form light, and store at –80 °C avoiding multiple freeze-thaw cycles.

14. Determine the protein concentration and degree of labeling by measuring the absorbance at 280 nm and at the wavelength corresponding to the absorbance maximum of the fluorescence dye (*see* **Note 18**).

3.3 Visualization of Ligand:Receptor Complex Internalization and Trafficking by HILO Microscopy (See Note 19)

1. Pre-warm an aliquot of imaging medium at 37 °C.

2. Dilute an aliquot of fluorescently labeled TSH to a final concentration of 6 μg/mL in imaging medium or imaging medium supplemented with 80 μM dynasore (to inhibit receptor internalization). Pre-warm at 37 °C.

3. Take the 6-well plates containing the primary thyroid cells out of the incubator (37 °C, 5 % CO_2) and replace the medium with 2 mL of either imaging medium or imaging medium with 80 μM dynasore. Put the plates back to the incubator for 20 min.

4. Stimulate the cells with fluorescently labeled TSH by replacing the medium with 2 mL of the pre-warmed solution containing fluorescently labeled TSH diluted in either imaging medium or imaging medium with dynasore and incubate for additional 5–30 min in the incubator (37 °C, 5 % CO_2).

5. Wash the cells with 2 mL of either imaging medium or imaging medium with dynasore. Incubate for 5 min in the incubator. Repeat this washing procedure two more times (*see* **Note 20**).

6. Transfer the coverslip from the 6-well plate to the cell imaging chamber. Seal in the coverslip by tightening the screw of the chamber, taking care not to break the coverslip.

7. Wash the coverslip once with 300 μL of imaging medium or imaging medium with dynasore.

8. Put one drop of immersion oil on to the objective of the TIRF microscope. Place the coverslip above the objective and raise the objective to bring the cells in focus.

9. Set the angle of the incident laser beam to a value that is slightly below the critical angle for reflection (*see* **Note 21**). This results in a thin illumination sheet that crosses the sample (HILO illumination) [11].

10. Acquire image sequences (Fig. 2a) (*see* **Note 22**).

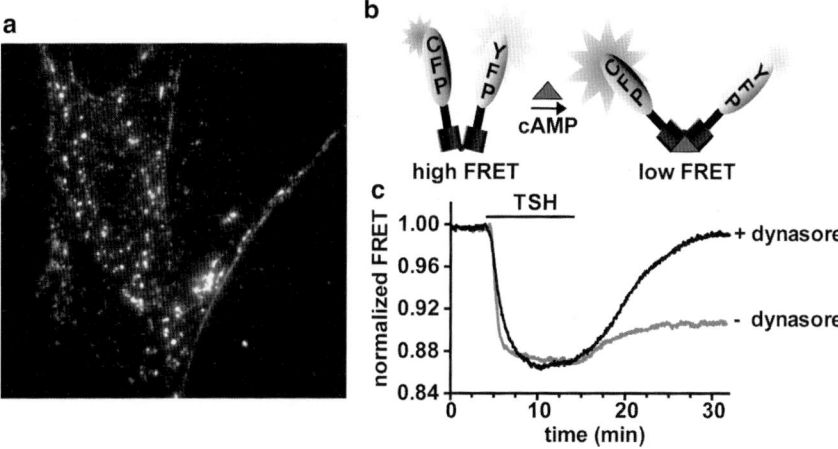

Fig. 2 Representative results. (**a**) A mouse primary thyroid cell was stimulated with Alexa Fluor 594-labeled TSH and visualized by HILO microscopy. The *bright spots* inside the cell represent TSH molecules that have been internalized into endosomes. (**b**) Schematic representation of the Epac1-camps sensor. Binding of cAMP leads to an increase of the distance between CFP and YFP, and hence to a decrease of FRET. (**c**) Real-time monitoring of cAMP levels by FRET microscopy in an intact thyroid follicle expressing the Epac1-camps sensor. A transient stimulation with TSH causes a largely irreversible decrease of FRET, i.e., an increase of cAMP levels. In contrast, pretreatment with dynasore to inhibit receptor internalization is associated with an almost fully reversible cAMP response (modified from ref. 2)

3.4 Mouse Primary Thyroid Cell Transfection by Electroporation	1. Pre-warm all solutions at 37 °C.

3.4 Mouse Primary Thyroid Cell Transfection by Electroporation

1. Pre-warm all solutions at 37 °C.

2. Aspirate medium from cells.

3. Wash cells twice with DPBS.

4. Incubate cells with 2 mL trypsin/EDTA solution for 10 min at 37 °C, 5 % CO_2 in an incubator.

5. During the incubation with trypsin/EDTA, wash 6 glass coverslips with DPBS, put them into the 6-well cell culture dish, and add 2 mL of complete culture medium in each well.

6. Stop the trypsin reaction by adding 8 mL complete culture medium to the cells and resuspend the cells by gently pipetting up and down 5–6 times with a glass pipette.

7. Transfer the cell suspension with a glass pipette to a 15-mL sterile plastic tube, tighten the cap, and centrifuge for 3 min at $300 \times g$.

8. Carefully aspirate the medium with a glass Pasteur pipette connected to an aspirator, add 10 mL DPBS, gently pipette up and down until the cells are resuspended, and centrifuge for 3 min at $300 \times g$ to completely remove the culture medium.

9. Aspirate the supernatant and resuspend the cells in 200 μL DPBS. Put 100 μL of the cell suspension in two separate electroporation cuvettes.

10. Add 20 μg of the Epac1-camps plasmid into each cuvette and gently mix.

11. Set the voltage and capacitance of the electoporator (*see* **Note 23**). Place one cuvette into the cuvette holder, close it, and discharge. Immediately remove the cuvette, add 1 mL of complete culture medium, and mix gently. Repeat this procedure for the second cuvette.

12. Add 350 μL of the cell suspension to each well of the 6-well plate containing the coverslips.

13. Incubate for 48 h at 37 °C, 5 % CO_2 in an incubator before use.

3.5 Real-Time Monitoring of cAMP Signaling by FRET Microscopy

1. Pre-warm an aliquot of imaging medium at 37 °C.

2. Dilute TSH to a final concentration of 30 U/L in imaging medium or imaging medium supplemented with 80 μM dynasore (to inhibit receptor internalization). Pre-warm at 37 °C.

3. Fill the perfusion system (*see* **Note 24**) with ultrapure water avoiding the formation of air bubbles. Wash the perfusion system by flushing it with ultrapure water for 30 min. Replace water with imaging medium and flush for additional 10 min. Remove the imaging medium and fill the different tubes of the perfusion system with the solutions required for the experiment, e.g., imaging medium, imaging medium + TSH, imaging medium + dynasore, imaging medium + dynasore + TSH.

4. Take a coverslip with thyroid cells transfected with the Epac1-camps sensor (Fig. 2b) (*see* **Note 25**) and place it in the imaging chamber. Seal in the coverslip by tightening the screw of the chamber, taking care not to break the coverslip.

5. Gently wash the coverslip thrice with 300 μL imaging medium. After the last wash, add 800 μL of imaging medium or imaging medium supplemented with dynasore and incubate for 30 min at 37 °C, 5 % CO_2 in an incubator.

6. Place the coverslip on the microscope stage and bring the cells in focus using the 10× objective.

7. Place the suction tubing connected to the aspirator on one side of the chamber just above the level of the liquid in the chamber.

8. Using the micromanipulator place the tip of the perfusion system on the opposite side of the suction tubing so that it is positioned at about two thirds of the distance between the center and the border of the field of view and slightly above the cells. Check that the perfusion system is working correctly using the same solution contained in the imaging chamber, i.e., imaging medium or imaging medium + dynasore. A slight movement of the cells should be observed when switching the flow on/off.

9. Change the objective to 63× without moving the sample, add a drop of immersion oil on the objective, and bring the cells in focus.

10. Turn on the fluorescence lamp and check CFP and YFP fluorescence. Find a suitable cell. Verify once more that the perfusion system is correctly working by switching the flow on and off. Switch on the perfusion and keep it on for the rest of the experiment. Switch on the aspirator.

11. Divert the light path to the camera port. Select the correct filters and illumination settings for CFP excitation. Take an image of the cell. Adjust acquisition parameters (exposure time, camera gain) and draw a region of interest (ROI) around the cell for the real-time calculation of FRET ratios.

12. Start the acquisition of a time series and wait for at least 10 min and until the baseline FRET signal is stable.

13. Stimulate with TSH for the desired time (e.g., 10 min) by switching the perfusion to the medium containing TSH.

14. Start washing by switching over to the medium without TSH and acquire images for at least 30 min.

15. Select the correct filters and illumination settings for YFP excitation and acquire an image.

16. Save the images and intensity values in the CFP and YFP channels.

17. At the end of the experiment, wash the perfusion system extensively by flushing it with ultrapure water. Store with ultrapure water inside until the next use.

18. Use appropriate software for performing FRET corrections (*see* **Note 26**) and visualizing the results (Fig. 2c).

4   Notes

1. To reduce background fluorescence and cell autofluorescence, we recommend using phenol red-free media for the culture of cells and the preparation of imaging medium.

2. We found that the back of a small plastic pipette tip is a good tool for spreading the collagen gel. To ensure uniform distribution and thus even thickness of the collagen gel, we start from the center of the objective and move centrifugally describing a spiral path with the back of the tip.

3. We use Vannas spring scissors with a cutting edge of 2.5–3 mm.

4. A similar procedure can be used to label other protein and peptide hormones, whereas various fluorescent small ligands, some of which are commercially available, have been synthesized.

5. We have successfully labeled TSH with the following dyes: Alexa Fluor 488, Alexa Fluor 594, and Alexa Fluor 647, giving green, red, and far-red fluorescence, respectively.

6. We use a 30-cm high, 1-cm wide glass column.

7. We recommend performing these experiments at 37 °C to better mimic physiological conditions. In addition, we recommend maintaining the optical components of the microscope, the imaging chamber, and all solutions at the same temperature. This is important to avoid focus drifts and similar artifacts during the course of the experiment. If using a microscope incubator, it is preferable to keep it always on, since it takes several hours for the microscope to equilibrate after turning the heating on. Moreover, temperature changes may induce vapor condensation inside the microscope.

8. The presence of BSA in the imaging medium reduces the binding of TSH to glass coverslips.

9. Epac1 is a guanine nucleotide exchange factor that contains a catalytic and a regulatory domain. It exists in an inactive state when the cAMP levels in the cell are low. When cAMP is produced, for example as a result of GPCR activation, it binds to the regulatory domain of Epac1 thus releasing it from the catalytic domain. This model of Epac1 activation in combination with fluorescence resonance energy transfer (FRET) has been exploited in designing a cAMP sensor, termed as Epac1-camps [10]. This sensor consists of the cAMP-binding domain of Epac1 sandwiched between the fluorescent proteins CFP and YFP. When the sensor is not bound by cAMP, there is FRET between CFP and YFP. When cAMP binds to the sensor, a conformational change occurs resulting in CFP and YFP to move away from each other, which is accompanied by a decrease of FRET. Thus, the degree of FRET measured in the presence of the Epac1-camps sensor is inversely related to the intracellular concentration of cAMP. The Epac1-camps sensor can be expressed in cells by transfection or other means to monitor cAMP levels in real time. A plasmid containing the Epac1-camps sensor is available upon execution of an MTA with the University of Würzburg.

10. This type of setup allows us to excite CFP and simultaneously acquire a CFP and YFP image of the sample, which are separated and projected to the left and right half of the camera by means of the beamsplitter. A raw FRET ratio image can then be calculated by dividing the intensity value in each pixel of the YFP image by the corresponding value in the CFP image. This, so-called ratiometric, FRET approach has the advantage of largely reducing the noise due to fluctuations of illumination intensity and camera sensitivity between different frames. Some software packages can perform this calculation online and thus allow one to follow the changes in FRET ratios during the course of an experiment. We recommend using a software with such a feature, as it facilitates the recognition of problems (e.g., focus instability, excessive bleaching) or artifacts during the

course of an experiment, which saves a considerable amount of time and reagents. Subsequent calculations, which can be performed off-line, are needed to obtain corrected FRET ratios.

11. The presence of BSA in the imaging medium reduces the binding of TSH to the perfusion system.

12. We sterilize the skin by briefly immersing the mouse in 100 % ethanol.

13. Be as gentle as possible to avoid damaging the thyroid tissue with the forceps.

14. Three mice are generally sufficient to prepare 6-9 coverslips with intact follicles or a confluent 10-cm Petri dish of primary thyroid cells.

15. Use of a water bath is preferred to that of a thermomixer because it ensures a more reliable temperature control. Avoid using a shaker or vortex as they cause damage to the follicles.

16. The packed column can be stored at 4 °C with approx. 2 mL of wash solution above the Sephadex G-25.

17. Incubating the solution on ice for about 20 min also helps in dissolving it completely.

18. Detailed instructions are contained in the manual available on the web site of the manufacturer of the reactive fluorescent dyes.

19. Different microscopy techniques, such as epifluorescence, confocal, or HILO microscopy, can be used to monitor ligand-receptor internalization and trafficking. We prefer HILO to other techniques because it allows rapid image acquisition, has a very high sensitivity, which is generally needed to work with endogenous receptors, and is particularly suited for primary thyroid cells, which spread on the coverslip surface and become rather thin compared to other cells.

20. If desired, cells can be fixed at this stage with 4 % paraformaldehyde in DPBS followed by three additional washes with DPBS.

21. The angle of the incident light beam needs to be determined empirically on a fluorescent sample. Begin with incident light perpendicular to the sample (i.e., epifluorescence). Then, increase the angle of the incident light until a TIRF image is produced. This is typically seen as a strong enhancement of the intensity of the objects (e.g., cell plasma membrane, dust particles on the glass surface) adjacent to or in direct contact with the glass surface and a disappearance of the fluorescence from the planes above. Once TIRF illumination is achieved, slightly decrease the TIRF angle until inner parts of the cell become visible, but without completely losing the enhancement of intensity and signal-to-noise ratio compared to epifluorescence illumination.

22. A number of parameters need to be defined for the acquisition of the image sequence. The most important are laser intensity, frame rate, exposure time, and EMCCD gain. High laser intensity, frame rate, and exposure time result in better images (i.e., high signal-to-noise ratio) but also in a faster light-induced destruction (photobleaching) of the fluorophores. Conversely, lower values for these parameters result in poorer image quality but allow extended observation of the sample. A relatively high on-chip gain can be used with EMCCD devices without introducing significant noise; however very high values can produce considerable noise. The best combination of these parameters should be determined experimentally to obtain a frame rate, an image quality, and an observation time sufficient to follow the phenomenon under investigation. We typically acquire images every 30 s for up to 60 min with an acceptable degree of photobleaching at the end of the experiment.

23. We use a Bio-Rad Gene Pulser electroporator with a Capacitance Extender set on 0.32 kV and 125 μF. An adjustment of these parameters might be required if using different equipment or cuvettes.

24. The perfusion system allows one to maintain the imaged cell under laminar flow and to rapidly change the solution that the cell is receiving. This allows us to apply controlled stimuli with an agonist and to rapidly and thoroughly wash away the agonist at the end of the stimulation.

25. Alternatively, we have used intact thyroid follicles isolated from transgenic mice with ubiquitous expression of the Epac1-camps sensor that were generated in our lab [2]. These mice show healthy development and a normal life expectancy as compared to wild-type mice, indicating that the expression of the sensor in most of the cells does not interfere with physiological processes. For more information regarding this sensor, refer to **Note 9**. We have used different primary cells (e.g., embryonic fibroblasts, cortical neurons, cardiomyocytes, and peritoneal macrophages) and more complex preparations (e.g., thyroid follicles and acute pituitary slices) obtained from these mice to monitor cAMP responses to different stimuli in real time [2, 8].

26. Corrected FRET ratios are calculated based on the measured background, CFP and YFP intensities, and knowledge of correction factors. These correction factors are dependent on the light source and set of filters used, and should be periodically determined experimentally for each microscopy setup. The first step consists in transfecting cells with either CFP or YFP and acquiring images of the transfected cells in both emission channels with the illumination conditions (filters, monochromator settings) used for either CFP or YFP excitation. From the intensity values obtained in the different images after subtracting the

background, the following correction factors are calculated. The first correction factor (A) expresses the fraction of the CFP intensity that is also seen in the YFP channel (bleedthrough) and can be calculated by dividing the intensity obtained with CFP-transfected cells imaged at CFP excitation (436 nm) in the YFP channel by that obtained in the CFP channel with the same illumination. The second factor (B) expresses the degree at which YFP is directly excited at the wavelength used for CFP excitation and can be calculated by dividing the intensity obtained with YFP-transfected cells imaged at CFP excitation (436 nm) in the YFP channel by that obtained in the same channel with YFP excitation (500 nm). Typical values are 0.5–0.6 for A and about 0.05 for B. Once these correction factors are known, the corrected FRET values can be calculated according to the following formula: $FRET_{corr} = YFP_{436} - A\ CFP_{436} - B\ YFP_{500}$, where YFP_{436} is the YFP intensity measured at a given frame of the FRET experiment (i.e., at CFP excitation = 436 nm), YFP_{500} is the YFP intensity acquired with YFP excitation at the end of the experiment (i.e., with 500-nm excitation), and A and B are the previously determined correction factors.

References

1. Lohse MJ, Nikolaev VO, Hein P et al (2008) Optical techniques to analyze real-time activation and signaling of G-protein-coupled receptors. Trends Pharmacol Sci 29:159–165

2. Calebiro D, Nikolaev VO, Gagliani MC et al (2009) Persistent cAMP-signals triggered by internalized G-protein-coupled receptors. PLoS Biol 7:e1000172

3. Müllershausen F, Zecri F, Cetin C et al (2009) Persistent signaling induced by FTY720-phosphate is mediated by internalized S1P1 receptors. Nat Chem Biol 5:428–434

4. Ferrandon S, Feinstein TN, Castro M et al (2009) Sustained cyclic AMP production by parathyroid hormone receptor endocytosis. Nat Chem Biol 5:734–742

5. Irannejad R, Tomshine JC, Tomshine JR et al (2013) Conformational biosensors reveal GPCR signalling from endosomes. Nature 495:534–538

6. Calebiro D, Nikolaev VO, Persani L, Lohse MJ (2010) Signaling by internalized G-protein-coupled receptors. Trends Pharmacol Sci 31:221–228

7. Lohse MJ, Calebiro D (2013) Cell biology: receptor signals come in waves. Nature 495:457–458

8. Jacobs S, Calebiro D, Nikolaev VO et al (2010) Real-time monitoring of somatostatin receptor-cAMP signaling in live pituitary. Endocrinology 151:4560–4565

9. Werthmann RC, Volpe S, Lohse MJ, Calebiro D (2012) Persistent cAMP signaling by internalized TSH receptors occurs in thyroid but not in HEK293 cells. FASEB J 26:2043–2048

10. Nikolaev VO, Bünemann M, Hein L et al (2004) Novel single chain cAMP sensors for receptor-induced signal propagation. J Biol Chem 279:37215–37218

11. Tokunaga M, Imamoto N, Sakata-Sogawa K (2008) Highly inclined thin illumination enables clear single-molecule imaging in cells. Nat Methods 5:159–161

INDEX

Bruce G. Allen and Terence E. Hébert (eds.), *Nuclear G-Protein Coupled Receptors*, Methods in Molecular Biology,
vol. 1234, DOI 10.1007/978-1-4939-1755-6, © Springer Science+Business Media New York 2015

Printed by Printforce, the Netherlands